Advances in Anatomy, Embryology and Cell Biology
Ergebnisse der Anatomie und Entwicklungsgeschichte
Revues d'anatomie et de morphologie expérimentale

48 · 2

A. Sousa-Pinto

Cortical Projections of the Medial Geniculate Body in the Cat

With 19 Figures

Springer-Verlag Berlin Heidelberg GmbH 1973

A. Sousa-Pinto, M. D., Ph. D.
Anatomical Institute
Medical Faculty
University of Oporto
Oporto (Portugal)

DOI 10.1007/978-3-642-95247-0

Inhalt

Einleitung 7

I. Das Glomus caroticum als chromaffines Paraganglion 7

II. Das Glomus caroticum als Chemoreceptor 8

III. Die Wahl des Untersuchungsobjektes 9

Material und Methoden 9

I. Lupenpräparation 10

II. Lichtmikroskopie 10

A. Fluorescenzmikroskopischer Nachweis von Catecholaminen . . . 10

B. Semidünnschnitte 10

III. Elektronenmikroskopie 12

A. Morphologische Untersuchung 12

B. Nachweis intravenös verabreichter Meerettichperoxydase 12

Befunde 12

I. Anatomische Präparation — Rekonstruktionen 12

A. Lupenpräparation 12

B. Serienschnitte und graphische Rekonstruktionen 16

II. Lichtmikroskopische Befunde an Semidünnschnitten 18

A. Das Glomusparenchym 18

B. Die Nerven des Glomus caroticum 19

C. Die Gefäße des Glomus caroticum 19

D. Die Bindegewebszellen 21

III. Fluorescenzmikroskopischer Nachweis von Catecholaminen 23

A. Das Glomusparenchym 23

B. Paraganglionäre Zellen in Truncus sympathicus 23

C. Mastzellen 26

D. Adrenerge Gefäßnerven 26

IV. Die Feinstruktur des Glomus caroticum 26

A. Das Glomusparenchym 26

1. Cytologie der chromaffinen Zellen (Glomuszellen, Typ I-Zellen) . 26

2. Die spezifischen Speichergranula 31

3. Die Stützzellen (Hüllzellen, Typ II-Zellen) 35

4. Die Nervenbeziehungen der chromaffinen Zellen 37

B. Die Nerven des Glomus caroticum 39

1. Nerven mit Perineurium 39

2. Das Verhalten des Perineuriums zum Glomusparenchym . . . 42

3. Nervenfasern ohne Perineurium 43

C. Die Gefäße des Glomus caroticum 43
1. Arteriolen 43
2. Capillaren 46
3. Venen 48
4. Arterienklappen 48
5. Gefäßnerven 48
D. Das Bindegewebe des Glomus caroticum 50
1. Fibrocyten 50
2. Makrophagen 52
3. Mastzellen 52
V. Markierung der Diffusionswege mit intravenös injizierter Meerrettichperoxydase 52
A. Lichtmikroskopischer Nachweis des Markierungsfermentes 54
B. Elektronenmikroskopischer Nachweis des Markierungsfermentes . 54
1. Der Transport durch Gefäßwände 54
2. Die Verteilung im Glomusparenchym 56
3. Das Perineurium als Diffusionsbarriere 56
4. Diffusion in den Endoneuralraum 56

Diskussion 58
I. Die Catecholamine im Glomus caroticum 58
A. Fluorscenzmikroskopischer — histochemischer Nachweis und chemische Analysen 58
B. Die Bedeutung der elektronenmikroskopisch nachweisbaren Speichergranula 61
1. Unbehandelte Tiere 61
2. Versuche mit Reserpin 63
3. Stimulierung des Glomus caroticum 64
4. Die Bedeutung der Speichergranula 64
II. Das Gefäßsystem des Glomus caroticum 66
III. Die Verteilung eines Markierungsproteins 67
IV. Das morphologische Äquivalent des Chemoreceptors 68
A. Erregung der afferenten Axonendigungen durch einen Transmitter 68
1. Acetylcholin als Transmitter 69
2. Catecholamin als Transmitter 69
B. Direkte Erregung markfreier Axonendigungen 70
C. Die Erregung des Receptors 70
D. Die Nervenbeziehungen der chromaffinen Zellen 71
1. Efferente Synapsen 71
2. Die Nervenversorgung des Glomus caroticum. 71
3. Afferente Synapsen 72
4. Receptorische Axonendigungen 72

Zusammenfassung 74
Summary 75
Literatur 77
Sachverzeichnis 83

Introduction

Monakow (1882) recognized in the medial geniculate body (MGB) a medial or "magnocellular", and a lateral, "principal" or parvicellular part. The latter was further subdivided by Cajal (1902) in a dorsal and a ventral lobe.

Monakow's division was adopted by Rioch (1929) and has been widely used. Morest (1964), in a Golgi study in the cat, used Cajal's scheme and further subdivided the ventral division into several subnuclei. Morest considered Cajal's suprageniculate nucleus as a part of the MGB.

The cortical projections of the MGB in the cat were studied by Woollard and Harpman (1939) and Ades (1941) with the Marchi method and, more recently, by Wilson and Gragg (1969) using the Nauta method. While all these studies show MGB projections to the first auditory area (AI) as electrophysiologically defined (Woolsey, 1960), there is disagreement as to the existence and extent of MGB projections to other cortical areas (AII, AIII, Ep, Ea and SF of Woolsey, 1960) where auditory activity has been electrophysiologically demonstrated. These studies did not disclose topical relationships between the different subdivisions of the principal MGB and the various cortical projection areas.

Rose and Woolsey (1949, 1958), Diamond *et al.* (1958), Neff and Diamond (1958), Locke (1961), and others studied atrophic changes and cell loss in the MGB after cortical lesions. These studies have shown that clearcut changes occur in the ventral division of the MGB when AI is destroyed and that extirpation of the other cortical auditory areas mentioned above only induces changes in the MGB when associated with extirpation of AI. In the latter experimental situation, changes in the MGB extend in some cases to the dorsal division of the MGB. As an attempt to explain these results Rose and Woosley (1949) set forth the hypothesis of "essential" and "sustaining" projections.

The anatomical studies mentioned above do not show the cortical projections of the MGB in great detail. To obtain precise information on the organization of this projection, necessary also for correct sampling in electron microscopic studies of the auditory cortex, it was decided to re-study the cortical projections of the principal MGB by means of silver impregnation methods. A particular aim of the present study has been to decide if the different subdivisions of the MGB have separate cortical projection areas, and if there is a topical organization within these projections, especially those to AI. If there are separate cortical projection areas for each subdivision of the MGB, it would be of interest to know their precise situation in the cortex.

Material and Methods

Altogether 21 adult cats were used for this investigation. The animals were operated upon under pentobarbital anaesthesia and with aseptic precautions. Stereotaxic lesions were produced by inserting an electrode, insulated except for the tip with Araldite. Through a

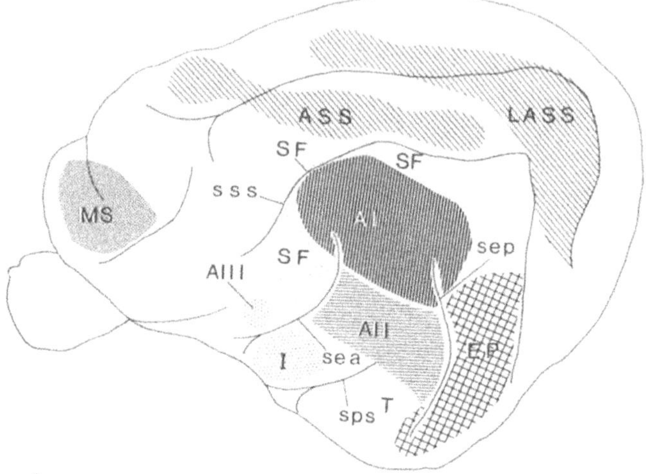

A

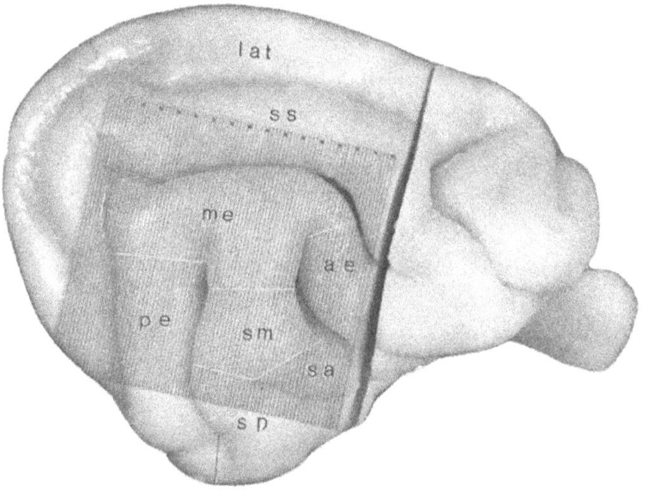

B

Fig. 1. A: Drawing of the cat cortical surface showing the auditory and other areas where auditory evoked activity has been recorded (modified from Woolsey, 1960). B: Photograph of the cat cortical surface with a superimposed grid (see explanation in the text). Labels of gyri as used in the present study. White lines show conventional boundaries between different parts of the gyri

small burr hole in the calvarium the electrode was passed obliquely at an angle of 35° to the horizontal plane and 15° to the sagital plane. The tip of the electrode was aimed at the Horsley-Clark coordinates between A6.5/L9.6/+10 and A8.5/L11.5/+11.5. An anodic current of 2.5 ma. was passed for 10 to 30 sec. In 4 animals the electrode tip did not reach the MGB. These animals were used for control of the degeneration caused by the passage of the electrode.

The animals were allowed to survive for 3 to 13 days. They were then killed by intravascular perfusion of 200 ml of 0.9% NaCl followed by 2 l. of 10% formol-saline. The perfusions were made through the abdominal aorta in the rostral direction. The brains were then quickly dissected out and fixation was continued by immersion of the brain in the same fixative. After 10 to 90 days, the rostral part of the ipsilateral hemisphere was separated by a transverse section passing rostral to the anterior ectosylvian gyrus (Fig. 1B). The lateral surface of the hemisphere was photographed and its posterior part was cut at 25 μm on the freezing microtome, in a direction parallel to the plane of separation between the rostral and caudal parts. The sections were collected in groups of 10, and kept a few days in 2% formalin.

One section from each group was stained according to the silver impregnation method of Wiitanen (1969) and counterstained with thionin. Another section of each group was stained with thionin, for the evaluation of the lesion. In most animals other sections of each group were stained according to the Nauta (1957), the Fink and Heimer (1967) or the Eager (1970) methods.

Lesions were studied in the thionin stained preparations. Sections containing the MGB were drawn with the aid of a projection apparatus and studied under the microscope. By comparison with corresponding normal myelin sheath and cell stained sections, and with the drawings and cytoarchitectonic description of the MGB given by Morest (1964), the extent of the lesion was evaluated and the destroyed parts of the nucleus were identified.

The sections containing the sylvian and ectoslyvian gryi were likewise drawn in the projection apparatus. The degeneration found in the microscopical study of these sections was entered in the drawings, terminal or pre-terminal degeneration as dots, passing fibres as wavy lines. Variations in the density of dotting give approximate indications of the intensity of the degeneration observed.

The drawings of the sections in which degeneration was entered were then compared with the photograph of the cortical surface mentioned above, using the same magnification for both. To provide accurate localization of each section in relation to the cortical surface, a grid of parallel lines, separated by a distance equivalent to the mean interval between two consecutive sections (250 μm) and corrected for the magnification used, was super-imposed on the photograph of the cortical surface (Fig. 1B). The degeneration entered in the drawings of the sections was then transferred to the corresponding strip of the cortical surface. The degeneration was carefully localized in this strip by measuring its distance to the suprasylvian sulcus. The final drawing of the cortical surface with the degeneration projected on it was then compared with that of the corresponding lesion, and with the cortical areas where auditory activity had been electrophysiologically recorded. Fig. 1a, shows the latter areas delimited according to Woolsey (1960), as well as the names of the relevant gyri and sulci as they will be used in the following presentation.

Results

Comments on the subdivision of the MGB

Taking advantage of the cytoarchitectonic details described by Cajal (1902) and Morest (1964), the different parts of the MGB and their subdivisions can be identified in Nissl stained preparations as stated by Morest (1965b), and their boundaries can be indicated. The larger cell bodies permit the separation of the magnocellular (medial) part of the medial geniculate body (MGB m). Morest's (1964) dorsal and ventral nuclei of the MGB can be identified on the basis of the density of the cells, and to a lesser extent according to their size. The cells are slightly larger in the dorsal nucleus (DO) and in the ventrolateral subdivision (VL) of the ventral nucleus. Cell bodies are slightly smaller and distinctly more densely packed (3:1.5 after Morest) in the pars lateralis (PL) and ovoidea (OV) of the ventral nucleus. This separates them from the cytoarchitectonically similar dorsal and ventrolateral nuclei. The pars lateralis and the pars ovoidea of the ventral nucleus can be separated from each other, due to their situation in relation to the myelinated fibres of the brachium of the inferior colliculus and

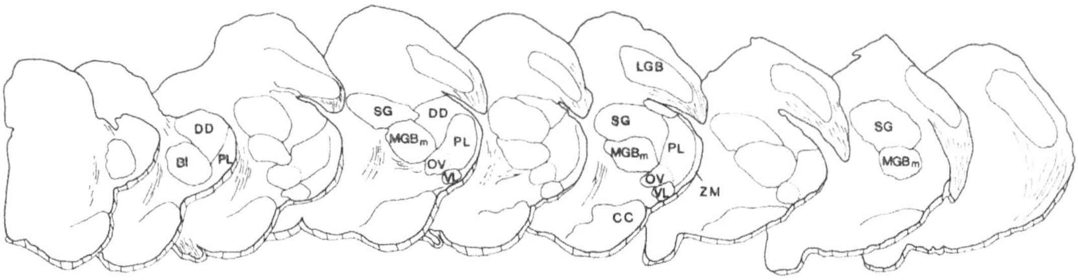

Fig. 2. Drawings of a series of equally spaced (250 μm interval) transverse sections through the medial geniculate body showing its subdivisions according to Morest (1964) and identified in the present study in Nissl stained preparations

to the somewhat more ordered arrangement of the cells in the pars lateralis. However, these two parts can not be sharply separated everywhere. The subdivisions of the MGB referred to above are shown in Fig. 2. The suprageniculate nucleus, included in the MGB by Morest but not by other authors (see Rinvik, 1968a, for a recent discussion of this point), as well as the magnocellular MGB were not included in this study, since their cortical projections have recently been investigated by Graybiel (1970, 1971, 1973) and Heath and Jones (1971).

Degenerating MGB efferent fibres: Morphology, time course and trajectory

Following lesions of the MGB (Fig. 3) it has been possible to stain degenerating fibres in the cortex when the animal survived 3 days or more.

Interrupted and varicose axons passing in the white matter or in the deep layers of the cortex, and forming organized bundles of fibres running in the same direction were interpreted as degenerating axons passing to other destination. Interrupted and irregularly swollen axons running in all directions between the cortical nerve cells were interpreted as terminal or pre-terminal degenerating fibres and a "field of termination" was recorded in the cortical areas where these were found. In these fields of termination small silver debris which, at least to some extent, may represent impregnated degenerating boutons, were frequently found, chiefly in the Wiitanen (1969) stained preparations.

Somewhat different observations were made in the parallel series of sections stained by the different silver impregnation methods. No substancial differences were found in the distribution of degeneration, but its density appeared consistently heavier in the Wiitanen (1969) and Eager (1970) stained preparations than in those impregnated by the Fink and Heimer (1967) and Nauta (1957) methods. Furthermore, although the degenerated axons in the white matter were similarly revealed by all 4 staining methods, when the Nauta method was used, degenerating fibres were found only in the deeper layers of the cortical areas of termination, while in adjacent sections stained by the Fink and Heimer method some degenerating fibres were found also in the IIIrd and IVth cortical layers, and in corresponding sections stained by the Wiitanen and the Eager methods very dense

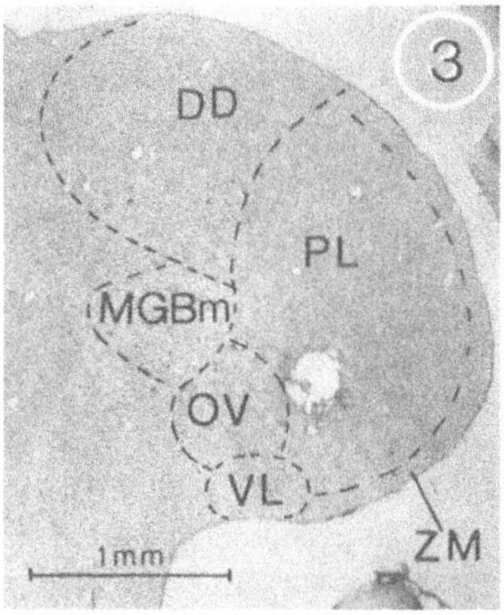

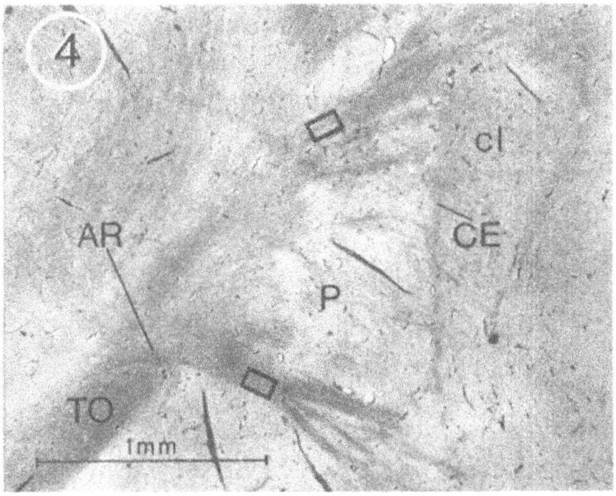

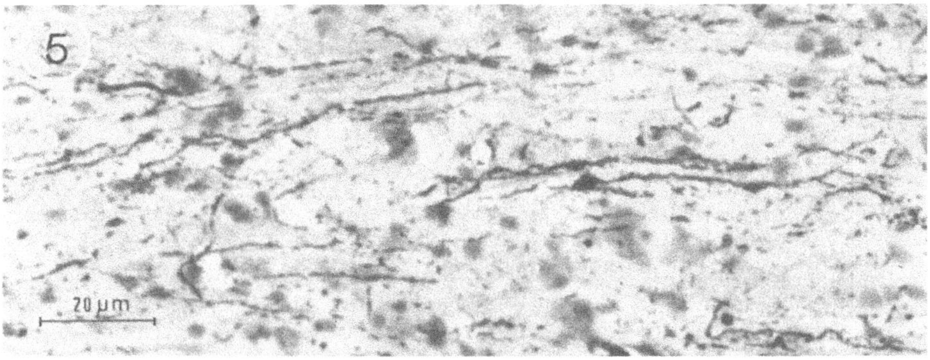

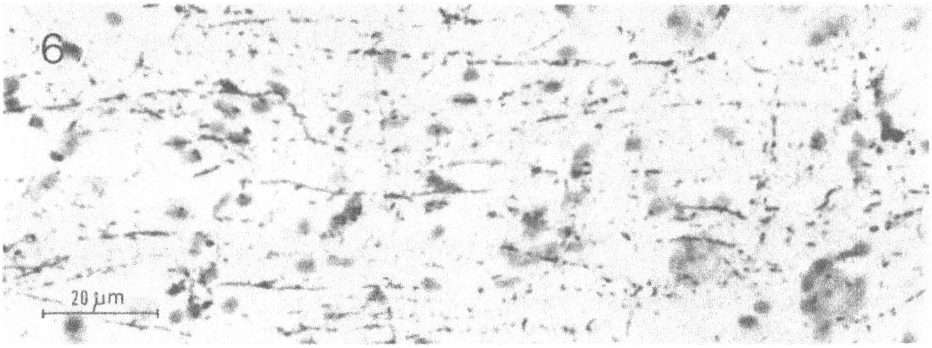

Fig. 3. Photograph of a thionin stained transverse section showing a lesion of the MGB and the subdivisions of the nucleus. Cat GSPCAU 44 (see section No. 29, Fig 14)

Fig. 4. Field photograph showing the putamen of cat GSPCAU 46 (see section No. 43, Fig. 17). The small frames show the places where the photomicrographs of Fig. 5 and 6 where obtained. Wiitanen (1969) method

Fig. 5. Degenerating geniculocortical fibres passing above the putamen at the place shown by the upper frame in Fig. 4

Fig. 6. Degenerating geniculocortical fibres passing below the putamen at the place shown by the lower frame in Fig. 4. Note that the fibres are thinner than those shown in Fig. 5

degeneration was found (Figs. 7 and 10) in all cortical layers except in the deep half of the Ist and in the IInd layers. Since in this laboratory the sections stained by the Eager (1970) method frequently showed very extensive silver precipitates, the following account is based chiefly on the preparations stained by the Wiitanen (1969) method.

In the animals which survived 3 (GSPCAU 21) or 4 (GSPCAU 25) days degeneration was scanty but its distribution was similar to that found in other animals with correspondingly situated lesions. However, in the Wiitanen (1969) sections of these two animals, the small silver droplets assumed to be degenerating boutons were not found. Most of the animals were killed 5, 6, 7 or 8 days after the lesion. In all of these, in Nauta (1957), Fink and Heimer (1968) and Wiitanen (1969) stained sections, degenerating fibres were abundant and had a similar morphology. Differences were found in the amount of argyrophilic pericellular small droplets (Fig. 10). In Wiitanen (1969) stained sections, these were abundant in animals with 7 days survival but were scanty in those with 5 or 13 days survival. In an animal killed 13 days after the operation (GSPCAU 30) thick degenerating fibres predominated, as compared with the others, and very scanty "bouton impregnation" was found. No impregnation of cortical cell bodies or dendrites similar to that ascribed by Grant and Aldskogius (1967) as being due to retrograde changes was found in any case.

In all experiments the degenerating fibres, leaving the site of the lesion, pass rostralwards along the medial surface and through the rostral pole of the MGB. They are collected in small bundles immediately rostral to the anterior tip of the MGB in the angular field situated medial to the optic tract and dorsal to the crus cerebri. Passing forwards, most of the fibres form a flattened ribbon lying close to the optic tract, which they cross rostrodorsally to enter the caudalmost part of the internal capsule and reach the medial border of the putamen. In their course until this point there are, apparently, fibres of all calibers among the MGB efferents. However, at the medial border of the putamen the degenerating fibres appear to split into two groups (Fig. 4).

The first group, containing fibres of medium and thick caliber, and occasionally very thick ones (Fig. 5), passes lateral and dorsalwards, across, dorsal or caudal to the putamen. Those passing across the putamen continue dorsalwards in the external capsule and join the fibres passing above that nucleus. Together, they pass dorsal and rostralwards around the dorsal border of the claustrum and, fanning out, they reach the cortex of the middle ectosylvian gyrus, the dorsal part of the posterior ectosylvian gyrus and the middle sylvian gyrus. The fibres to the most dorsal cortical areas pass upwards immediately underneath the cortex of the ventral

Fig. 7. Degenerating fibres in layer IV of AI in cat GSPCAU 46 (see section No. 48, Fig. 17). Wiitanen (1969) method

Fig. 8. Degenerating fibres in the temporal cortex of cat GSPCAU 46 (section see No. 33, Fig. 17). The magnification is the same as in Fig. 7. Wiitanen (1969) method

Fig. 9. Degenerating fibres passing to layer I through the lower cortical layers at the bottom of the dorsal part of the posterior ectosylvian sulcus. Cat CSPCAU 47 (see section No. 39, Fig. 16). Wiitanen (1969) method

Fig. 10. Impregnation of small argyrophilic globular particles, probably degenerating terminal boutons, AI of cat GSPCAU 45. Wiitanen (1969) method

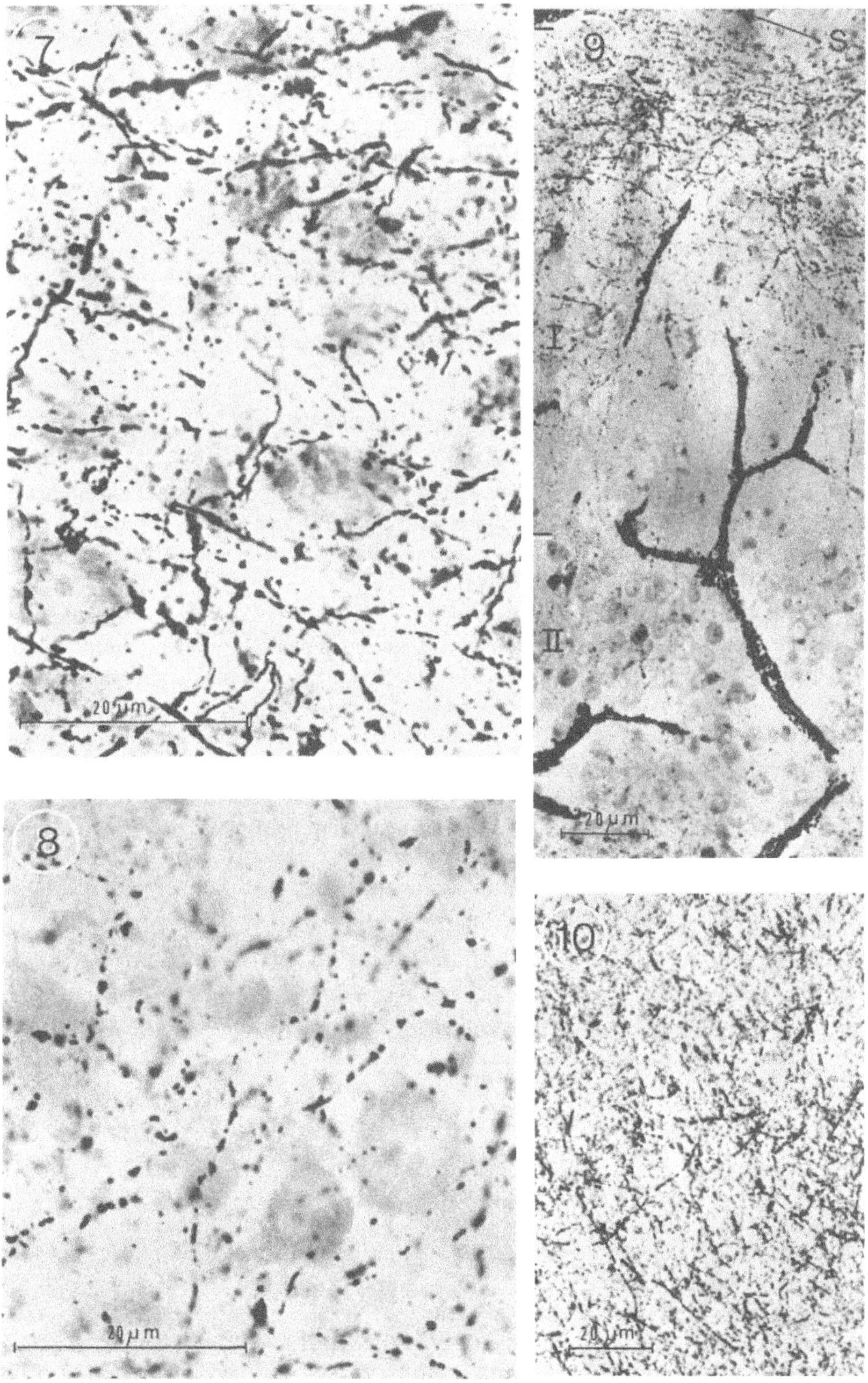

Figs. 7–10

areas. Reaching the deep cortical surface the fibres pass obliquely through the deep layers and seem to end around the cells of layers III and IV (Fig. 7). In most of the experiments a fair number of degenerating fibres cross the whole cortex at the bottom of the dorsal part of the posterior ectosylvian sulcus and reach layer I. These fibres spread tangentially in layer I along the banks of the sulcus, and reach the convexity of the neighbouring gyri (Fig. 9).

The second group of fibres arches lateralwards below the putamen (Fig. 4) and reaches the cortex of the ventral part of the posterior ectosylvian gyrus and the posterior sylvian gyrus as well as the corresponding banks of the posterior ectosylvian sulcus. The fibres of this group are thin, some of them very thin (Fig. 6). They run with no definite orientation between the cells of the deeper layers (IV, V and VI) of the cortical areas mentioned (Fig. 8) where there frequently is silver debris interpreted as bouton impregnation.

Presentation of Experiments

A. Fibre Degeneration due to the Electrode Track

Cat GSPCAU 34, 8 Days Survival (Fig. 11). The electrode passed through the cortex and white matter of the ventrocaudal part of the lateral gyrus as well as through the caudal hippocampus. Its tip reached the postsubicular cortex immediately caudal and lateral to the MGB.

The degeneration in the cortex of the medial surface of the hemisphere was not mapped. On the lateral surface, degeneration was found in the posterior part of the lateral gyrus around the electrode track and in the descending posterior part and caudal $^2/_3$ of the convexity of the middle suprasylvian gyrus. Degeneration was densest in the middle part of the middle suprasylvian gyrus, where it extended to the dorsal bank and bottom of the suprasylvian sulcus (Fig. 11, section No. 30).

In three other cats (GSPCAU 6, 10 and 12, killed, respectively, 5, 3 and 9 days after the operation) similar electrode tracks were found. The fibre degeneration found in the hemisphere of these animals was essentially as that in cat GSPCAU 34.

Comments. In these 4 animals the electrode track was similar to that used in the following experiments, but there was no lesion of the MGB. Cortical degeneration was found in several cortical regions, but *the auditory cortical areas were entirely free of degeneration. Therefore, in the following experiments with lesions of the MGB, no mention will be made of the degeneration found in the areas where it was present in this group of experiments, since the electrode was passed through the same area.*

B. Lesions of the Dorsal Nucleus of the MGB

Cat GSPCAU 9, 7 Days Survival (Fig. 12). The lesion was restricted to the dorsal nucleus of the MGB and destroyed it at caudal levels. The cortical degeneration was restricted to the middle part of the sylvian gyrus, where it was heavy. It was absent in the ectosylvian gyrus as well as in other parts of the sylvian gyrus. No degeneration was present in cortical layer I.

Cat GSPCAU 16,5 Days Survival (Fig. 12). The rather large lesion of the MGB destroyed extensively its dorsal nucleus. In addition, the lesion extended to the

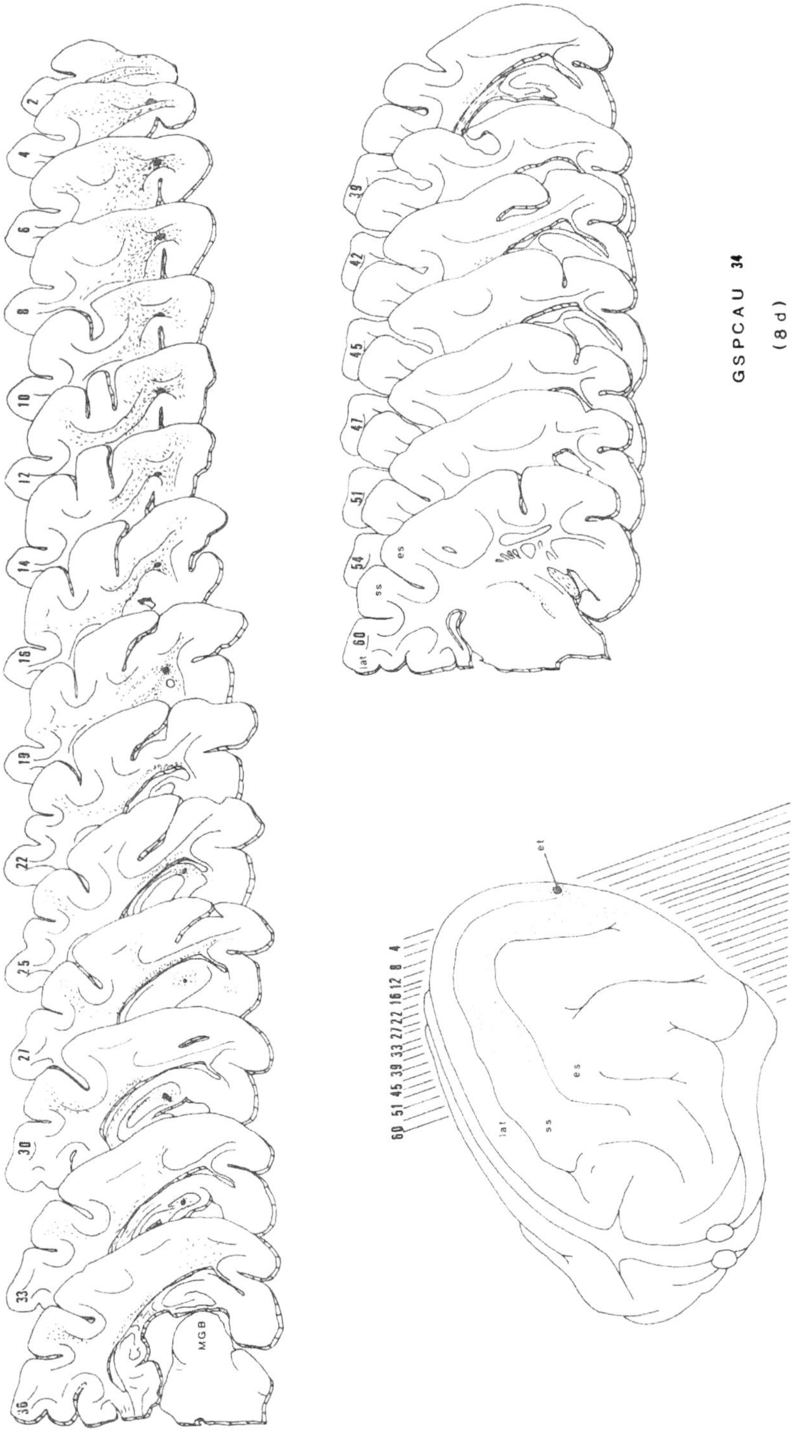

Fig. 11. Drawings showing the distribution of degeneration as mapped on the lateral surface of the cortex in cat GSPCAU 34, where the electrode did not reach the MGB

medial magnocellular MGB and, most likely, involved also the lowest part of the suprageniculate nucleus, as well as a very restricted area of the dorsomedial border of the ventral nucleus of the MGB.

The cerebral cortex presented slight deviations from the usual sulcal pattern (Fig. 12). Cortical degeneration was found in the sylvian gyrus and was also present in the anterior suprasylvian, anterior ectosylvian and anterior sylvian gyrus. Scanty degeneration was also found in the posterior sylvian gyrus. The middle and posterior ectosylvian gyrus were free of degeneration. No degeneration was found in cortical layer I.

Cat GSPCAU 25, 4 Days Survival (Fig. 12). The lesion and the distribution of cortical degeneration found in this animal were similar to those described above in cat GSPCAU 9. However, the lesion was smaller and degeneration was much lighter.

Comments. In these three experiments parts of the dorsal nucleus of the MGB were destroyed. In the cortex, degeneration was always present in the middle sylvian gyrus, corresponding to AII (Fig. 1).

In the second experiment, the medial MGB and adjacent areas of the suprageniculate nucleus and of the ventral MGB nucleus were damaged together with the dorsal nucleus. Consequently, in this experiment the cortical degeneration was more extensive than in the other two. It extended to insulo-temporal auditory areas, to AIII, to the ectosylvian anterior and rostral suprasylvian cortex. These cortical areas are the target of the magnocellular MGB as well as of the suprageniculate nucleus (Graybiel, 1970, 1971, 1973; Heath and Jones, 1971). Therefore, the lesion of those thalamic parts in this case is most likely responsible for the degeneration found in these cortical areas while the associated lesion of the dorsal nucleus of the MGB caused the degeneration in the middle sylvian gyrus. In all three experiments, AI and the ectosylvian posterior gyrus were free of degeneration. Therefore, it appears from these experiments that the *dorsal nucleus of the MGB projects exclusively to the middle sylvian gyrus, which coincides with most of the second auditory area (AII).* However, since no lesions of the most rostral parts of the dorsal nucleus were obtained, it is not known whether also these project to AII.

C. Lesions of the Lateral Part of the Ventral Nucleus

Cat GSPCAU 22, 7 Days Survival (Fig. 13). The lesion is confined to the lateral part of the ventral nucleus.

In the cortex heavy degeneration was restricted to the middle ectosylvian gyrus rostral to the tip of the posterior ectosylvian sulcus. The degeneration was densest in the middle and ventrocaudal parts of this area. No degeneration was found in cortical layer I and it was also absent in all layers of the other parts of the sylvian and ectosylvian gyri.

Cat GSPCAU 30, 13 Days Survival (Fig. 13). A restricted dorso-rostro-lateral area of the pars lateralis of the ventral nucleus was destroyed. In addition, the optic tract was extensively damaged.

In the cortical parts studied, degeneration was present exclusively rostrodorsally in the middle ectosylvian gyrus. There was no degeneration in layer I.

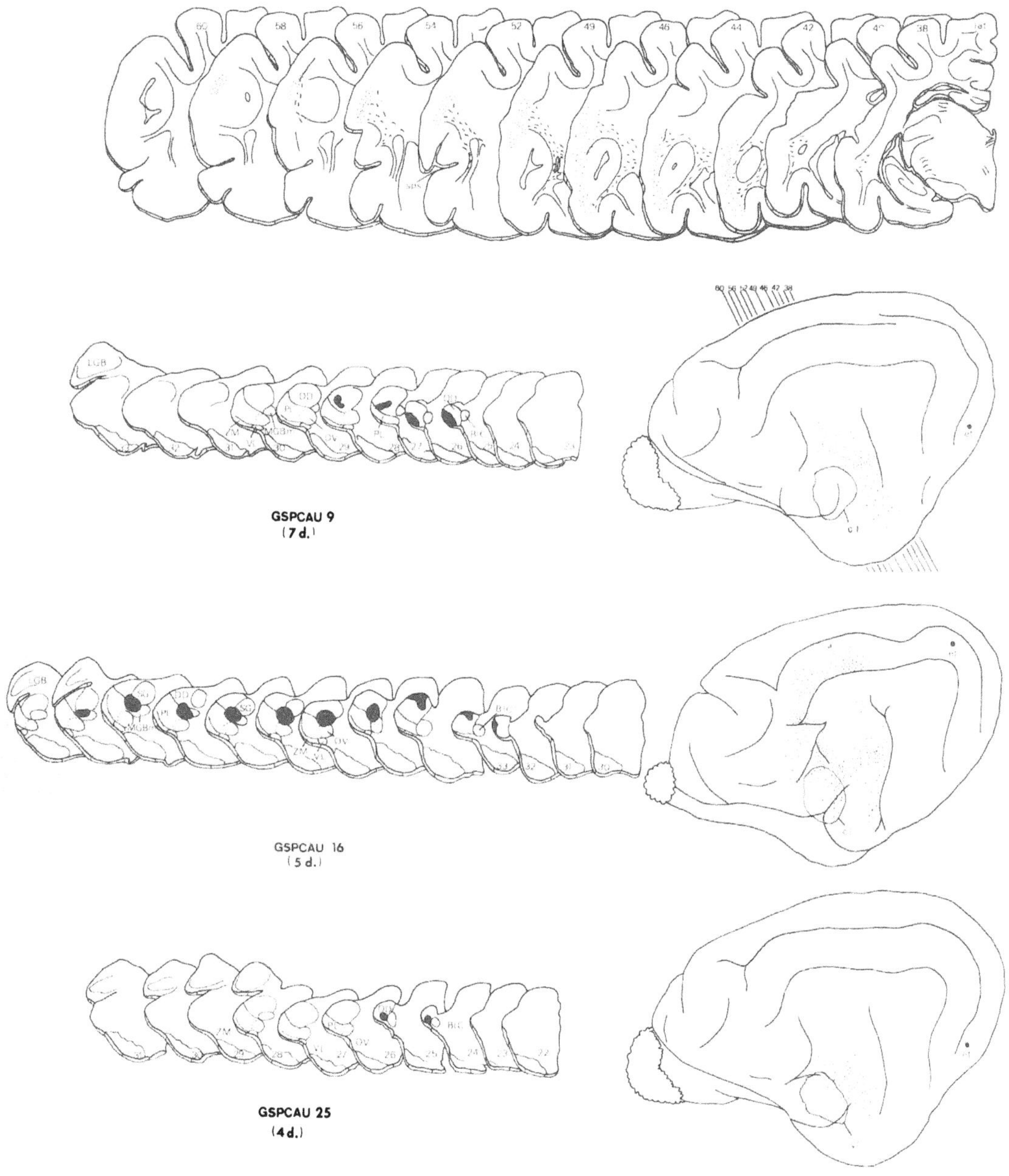

Fig. 12. Drawings showing the distribution of degeneration in the animals of group B. In this and in the following figures the drawings of serial transverse sections of the MGB to the left show the lesion in the case whose cortical surface is shown to the right at the same level. The serial cortical sections drawn on top of the figure belong to the animal whose brain is represented immediately below. The position of each particular section shown above is indicated by oblique lines in the drawing of the brain. The lesions are shown in black and the degeneration by dotting

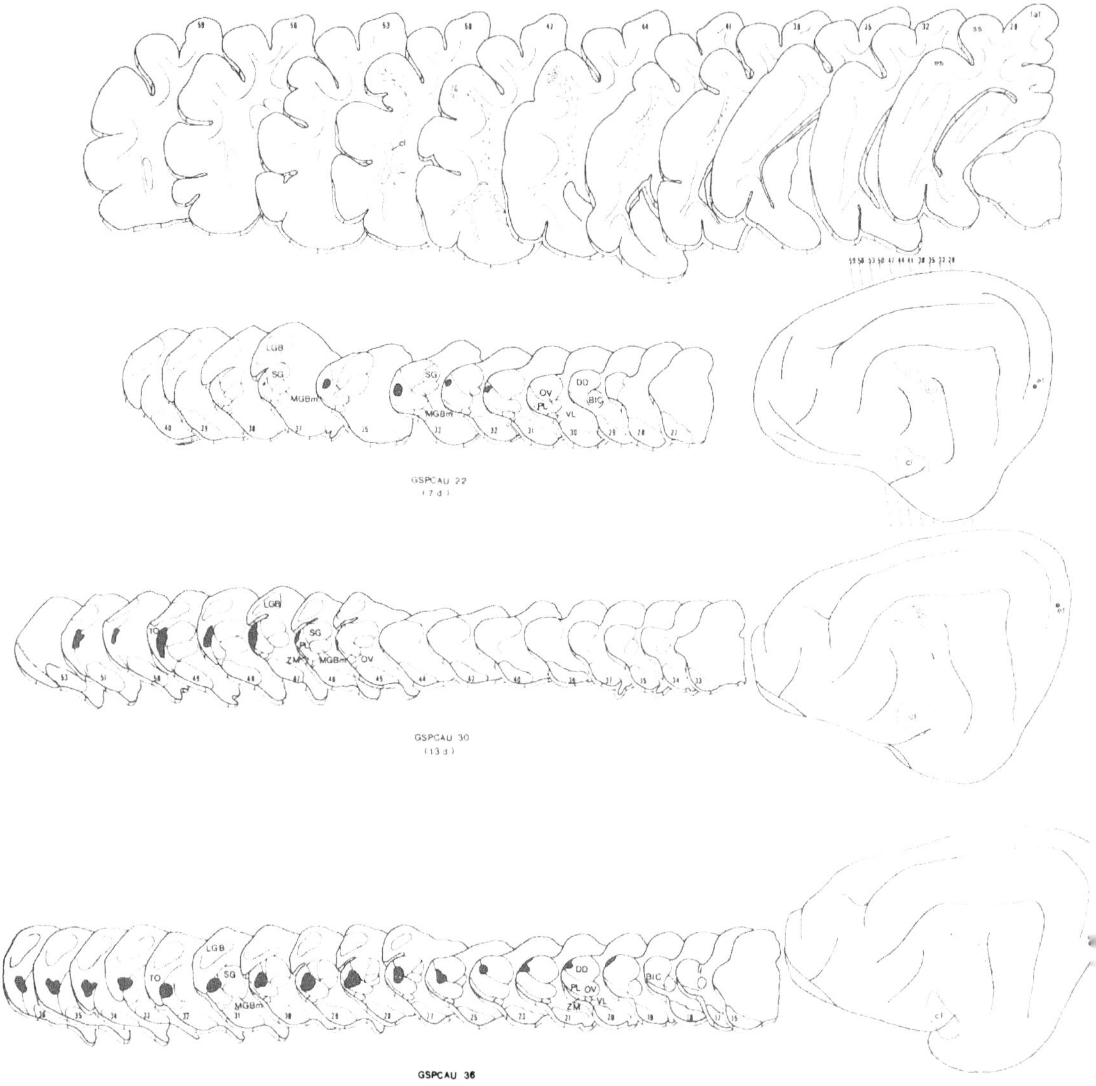

Fig. 13. Drawings showing the degeneration in the experiments of group C

Cat GSPCAU 36, 5 Days Survival (Fig. 13). The lesion was rather large. However, in the parvicellular MGB it was almost restricted to the lateral part of the ventral nucleus. Most of this subdivision was destroyed along its entire rostrocaudal extent. Caudally, there was a rather restricted involvement of the dorsal nucleus, and rostrally there was a small lesion of the thalamic reticular complex.

The cortical degeneration was very heavy and extended through all of the middle ectosylvian gyrus rostral to the tip of the ectosylvian sulcus. There was as small patch of degeneration in the middle sylvian gyrus immediately dorsal to the tip of the pseudosylvian sulcus. There was some degeneration in the cortex of the rostral bank and bottom of the dorsal part of the ectosylvian posterior sulcus. At this point fibres passing through the cortex proceeded rostralwards in layer I of the ectosylvian gyrus.

Comments. In this group, the lesions were mostly restricted to the lateral part of the ventral nucleus of the MGB. The ensuing cortical degeneration was restricted to AI. The degeneration following the extensive lesion in cat GSPCAU 36, as well as the findings in the two cases with more restricted lesions, show that the lateral part of the ventral nucleus of the MGB projects to the whole extent of AI. The small patch of degeneration in the middle sylvian gyrus in cat GSPCAU 36 can be explained by the associated small lesion of the dorsal nucleus (cp. the experiments of group B).

A comparison between the first and the second experiment in this group suggests that there is a topical organization of the cortical projection of this nuclear subdivision. The lesion in the first animal is more caudally situated than that in the second. The cortical degeneration in the first animal was found in a situation more ventral and caudal than that in the second (Fig. 13). The above suggestion is further supported by the analysis of all the experiments (see below and Fig. 18).

It is worth noting that in these 3 experiments there was no degeneration in the caudal bank of the dorsal part of the posterior ectosylvian sulcus and in the dorsal part of the posterior ectosylvian gyrus. These areas, where Rose and Woolsey (1949) found apical cochlear representation, appear to receive no fibres from the lateral part of the ventral nucleus of the MGB.

D. Lesions Involving the Pars Ovoidea of the Ventral Nucleus

Cat GSPCAU 38, 7 Days Survival. In the MGB the lesion was situated at the border between the dorsal and the ventral nuclei and destroyed both subdivisions extensively. The lesion of the ventral nucleus extended medially and seemed to cross the undefined border towards the pars ovoidea and to destroy a small part of the latter.

Heavy cortical degeneration was present in the middle ectosylvian and sylvian gyri and, dorsally, in the rostral part of the posterior ectosylvian gyrus. Lighter degeneration was found in the posterior sylvian gyrus, as well as in the whole dorso-ventral extent of the banks and bottom of the ectosylvian posterior sulcus. In the latter situation degeneration was present in layer I.

Cat GSPCAU 37, 6 Days Survival (Fig. 14, below). An extensive lesion destroyed chiefly the pars lateralis of the ventral nucleus, but extended medially beyond its border with the pars ovoidea. Caudally, a large area of the dorsal nucleus was destroyed.

In the cerebral cortex degeneration was heavy ventrally in the middle ectosylvian gyrus, rostral to the posterior ectosylvian sulcus, and in the middle and posterior sylvian gyrus. Degeneration was likewise heavy in the bottom and banks of the posterior ectosylvian sulcus and in the adjoining part of the convexity of the posterior ectosylvian gyrus, where it was most heavy rostrodorsally. There was degeneration in layer I.

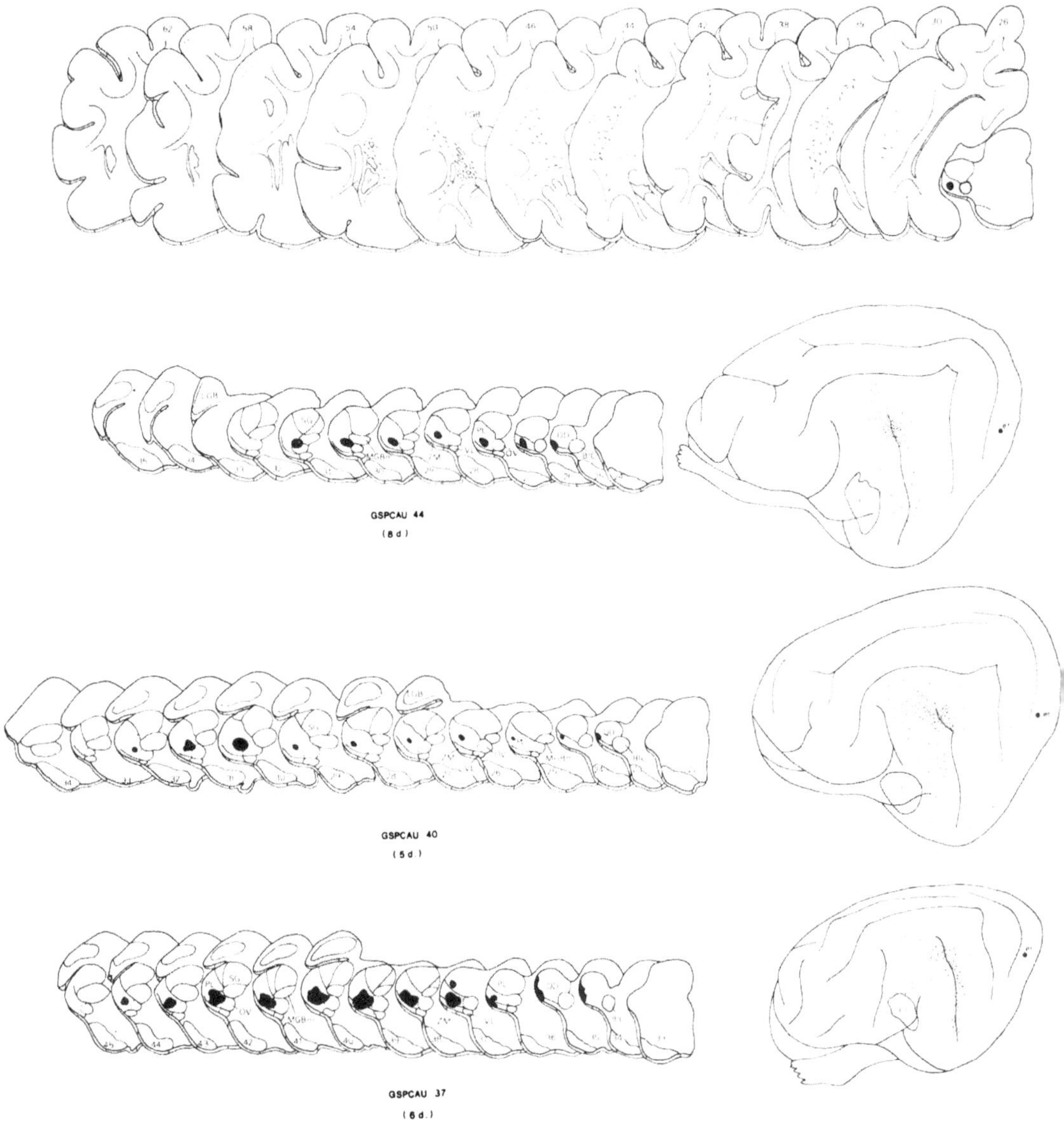

Fig. 14. Drawings showing the degeneration in the experiments of group D

Cat GSPCAU 44, 8 Days Survival (Fig. 14). In addition to an extensive area of the lateral part of the ventral nucleus, the lesion destroyed small areas of the pars ovoidea and, caudally, of the dorsal nucleus.

In the cortex, degeneration was present in the rostral parts of the middle and posterior ectosylvian gyri, in the middle and posterior sylvian gyri and in the banks and bottom of the posterior ectosylvian sulcus. There was heavy degeneration rostrally and caudally to the dorsal part of the posterior ectosylvian sulcus, where degeneration was also present in layer I.

Cat GSPCAU 40, 5 Days Survival (Fig. 14). A relatively small lesion extended through the pars lateralis and pars ovoidea of the ventral nucleus. Caudally, it involved in addition a very small area of the dorsal nucleus.

In the cortex degeneration was heavy in the areas immediately rostral and caudal to the dorsal tip of the posterior ectosylvian sulcus, as well as in the bottom and banks of that sulcus, where degeneration was found passing to layer I. In addition, scanty degeneration was found in the posterior sylvian gyrus and dense degeneration in the rostral part of the ventral posterior ectosylvian gyrus.

Comments. In these 4 experiments the lesions had destroyed parts not only of the dorsal nucleus but also of the pars lateralis of the ventral nucleus. In agreement with the experiments of the two preceeding groups there was degeneration in AI and AII. Furthermore, even if the border between pars lateralis and ovoidea of the ventral nucleus is not very well defined in these preparations, it appears that in all 4 experiments the pars ovoidea was affected as well. In the cortex, additional degeneration was found in the rostral half of the posterior ectosylvian gyrus and in the posterior sylvian gyrus, as well as in the bottom and banks of the ectosylvian sulcus. The degeneration in these areas differed in density except in the bottom and banks and immediately caudal to the dorsal part of the posterior ectosylvian sulcus where heavy degeneration was found in all 4 experiments. *This suggests that the latter small area is a preferencial projection area of the pars ovoidea of the ventral nucleus.* However, this nuclear subdivision appears to project also to the whole extent of both banks of the ectosylvian posterior sulcus as well as to the adjoining cortical surface.

E. Lesions Involving the Ventrolateral Nucleus

Cat GSPCAU 43, 6 Days Survival (Fig. 15). Almost the entire ventrolateral nucleus was destroyed together with the neighbouring ventral part of the pars ovoidea. Two patches of heavy degeneration were found in the cortex, one in the dorsal part of the posterior sylvian gyrus, the other immediately caudal to the dorsal part of the posterior ectosylvian sulcus. In addition, scanty degeneration was present rostrally in the middle and posterior ectosylvian gyrus and in the middle sylvian gyrus. Degeneration was present in cortical layer I, in the bottom and banks of the posterior ectosylvian sulcus.

Cat GSPCAU 48, 7 Days Survival (Fig. 15). In the MGB, an extensive area of the pars lateralis and ovoidea of the ventral nucleus was destroyed together with the rostral part of the ventrolateral nucleus.

In the cortex degeneration was essentially similar to that described in experiment GSPCAU 43. Heavy degeneration was present in the posterior sylvian gyrus and in the cortex immediately caudal to the dorsal part of the posterior ectosylvian sulcus. Degeneration was also dense in the rostral middle ectosylvian gyrus. There was degeneration in layer I, close to the dorsal part of the posterior ectosylvian sulcus.

Cat GSPCAU 42, 8 Days Survival (Fig. 15). Caudally, the lesion destroyed the pars lateralis and ovoidea of the ventral nucleus and rostrally a small area of the ventrolateral nucleus.

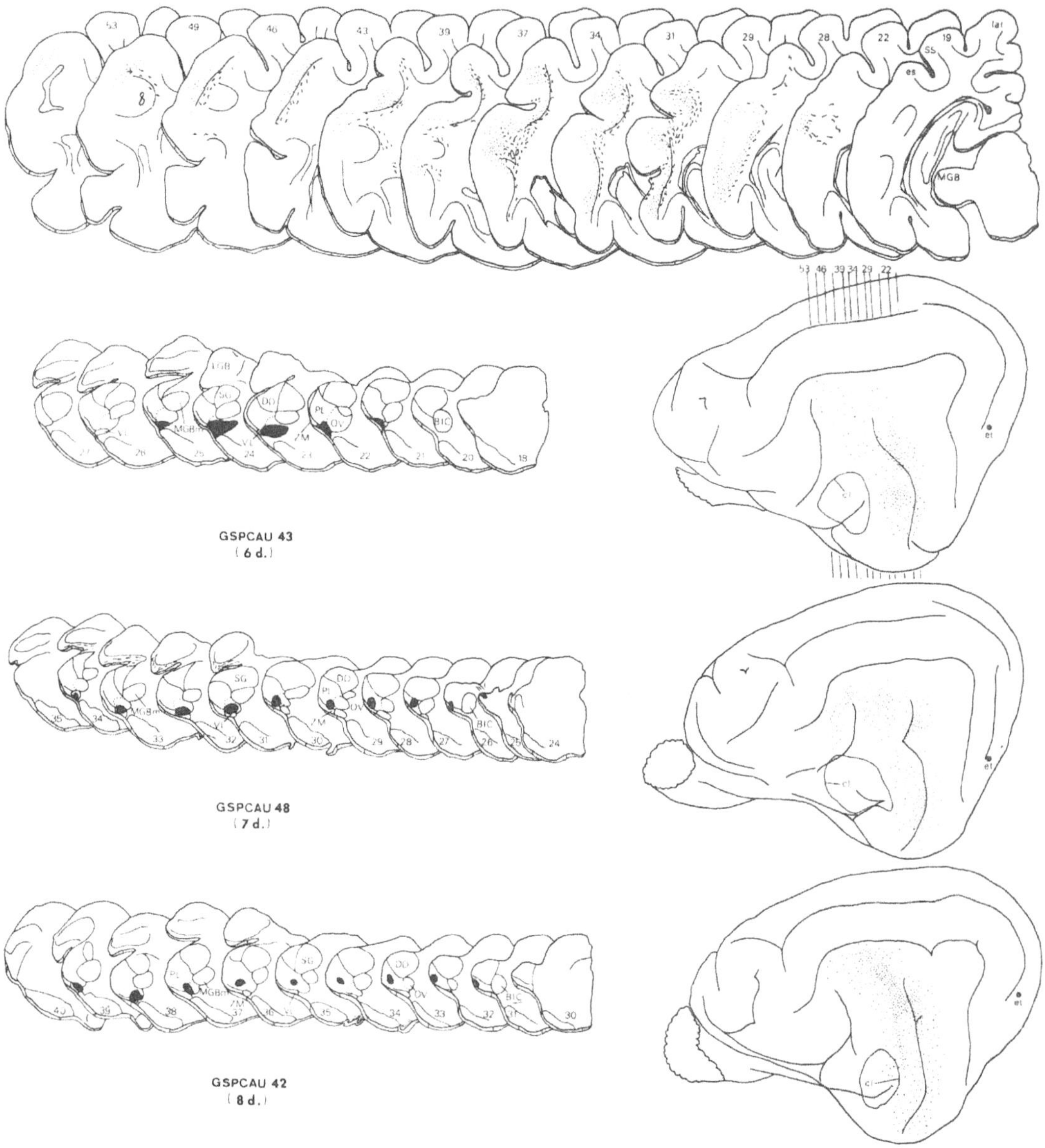

Fig. 15. Drawings showing the degeneration in the experiments of group E

In the cortex degeneration was found in the same areas as in the two preceeding experiments. However, in this case degeneration was lighter in the posterior sylvian gyrus, while it was much heavier and more extensive in the rostral part of the middle ectosylvian gyrus.

Comments. The lesions of this group destroyed parts of varying extention of the pars lateralis and ovoidea of the ventral nucleus while the dorsal nucleus was not significantly involved. Degeneration was found in the rostral middle ectosylvian gyrus and immediately caudal to the dorsal part of the posterior ectosylvian sulcus, while it was scanty in the middle sylvian gyrus. These results confirm the observations made in the series of experiments B, C and D.

Furthermore, in these cases the ventrolateral nucleus of the ventral MGB was involved as well in the lesions, although in GSPCAU 42 only a very small part of it was destroyed. In the first two experiments a patch of heavy degeneration was found in the posterior sylvian gyrus which was not present in the experiments of the former series with no lesion of the ventrolateral nucleus. The degeneration in this area was lighter in the third experiment, in which the lesion of that nucleus was smaller.

These observations suggest that the ventrolateral nucleus of the ventral division projects preferentially to the posterior sylvian gyrus. This area is, apparently, included in Ep (Fig. 1 A).

F. Lesions Involving the Marginal Zone

Cat GSPCAU 47, 6 Days Survival (Fig. 16). The lesion destroyed extensively the marginal zone of the MGB and a small area of the neighbouring pars lateralis.

A patch of very heavy degeneration was found in the ventral part of the posterior ectosylvian gyrus, immediately caudal to the ventral part of the posterior ectosylvian sulcus. Very heavy degeneration was likewise present in the bottom and banks of the dorsal part of the posterior ectosylvian sulcus—which was slightly aberrant in this animal—as well as in the neighbouring cortical surface, both rostrally and caudally. In this area degeneration was found in cortical layer I.

Cat GSPCAU 45, 8 Days Survival (Fig. 16 below). The lesion destroyed extensively the pars lateralis of the ventral nucleus and also the neighbouring marginal zone. There was a very small lesion of the dorsal nucleus.

Very heavy degeneration was found in the bottom and rostral bank of the dorsal part of the posterior ectosylvian sulcus as well as in the anterior half of the middle ectosylvian gyrus. In this area degeneration was found also in layer I. Lighter degeneration was present along the whole dorsoventral extent of the anterior part of the posterior ectosylvian gyrus.

Cat GSPCAU 21, 3 Days Survival (Fig. 16). The small lesion destroyed the marginal zone and a very small area of the neighbouring pars lateralis.

Cortical degeneration was scanty and its situation is difficult to compare with other experiments due to variation of the sulcal pattern. The posterior ectosylvian sulcus in this brain was nearly horizonal and was absent ventrally. Denser degeneration was found in a cortical area, which apparently corresponds to the rostral part of the middle ectosylvian gyrus. In the cortical area where the ventral part of the posterior ectosylvian sulcus is usually present very dense degeneration was found. No degeneration was found in layer I.

Comments. In this group of experiments the lesions destroyed mainly the marginal zone, together with small areas of the pars lateralis of the ventral nucleus. In the cortex, in addition to the degeneration found in the rostral middle

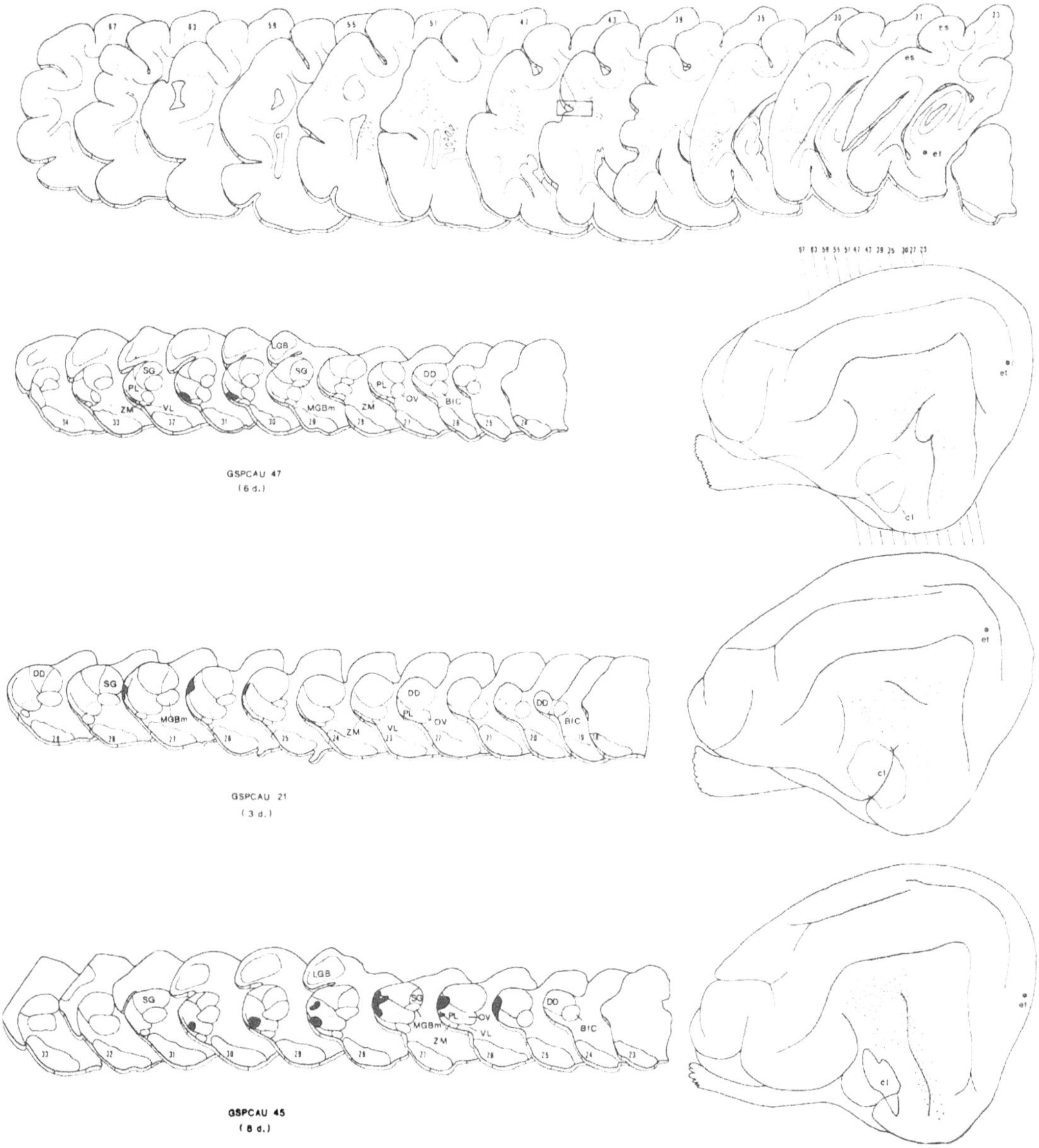

Fig. 16. Drawings showing the degeneration in the experiments of group F. The frame in the cortical section No. 39 shows the place where the photomicrograph of Fig. 9 was obtained

ectosylvian gyrus, which in the light of the previous experiments is due to the lesion of the pars lateralis, dense degeneration was found ventrally. In one case (GSPCAU 47) dense degeneration was present in the rostral part of the ventral posterior ectosylvian gyrus. In GSPCAU 21 its localizations appears to be the same, but the absence of the ventral part of the posterior ectosylvian sulcus does not allow a precise localization. In cat GSPCAU 45 the lesion of the marginal zone is small and degeneration in this area is not very dense, but it has the same localization as seen in cat 47. These findings seem to suggest *that the marginal zone may project preferentially to a cortical area located in the rostroventral part of the posterior ectosylvian gyrus.*

G. Lesion Involving the MGB Efferents

Cat GSPCAU 46, 6 Days Survival (Fig. 17). The stereotaxic lesion in this case destroyed a small caudal area of the dorsal nucleus and, to a large extent, both pars lateralis and ovoidea of the ventral nucleus rostrally. Furthermore, rostral to the principal MGB it destroyed its efferents, dorsal to the crus cerebri.

In the cerebral cortex, degeneration was rather heavy and extensive. It was present in the rostral half of the middle ectosylvian gyrus, along the whole dorsoventral extent of the rostral part of the posterior ectosylvian gyrus and in the bottom and banks of the posterior ectosylvian sulcus. Less heavy degeneration was found in the middle and posterior sylvian gyri. Degeneration was present in the cortical layer I around the dorsal tip of the posterior ectosylvian sulcus.

Comments. In this case the nuclear lesion was relatively large and, in addition, the MGB efferent fibres were extensively destroyed. Heavy cortical degeneration was found in all the particular areas where it was found following more or less restricted lesions in the previously presented experiments. Thus, *this experiment appears to demonstrate the total cortical projection area of the parvicellular MGB (Fig. 17).*

Summary of the Results

Summing up the results obtained in these 7 groups of experiments the following is apparent:

Group G, together with all the others, shows the entire cortical projection area of the principal MGB. This area covers (Fig. 17) AI, AII, and Ep of Woolsey (1960), the bottom and both banks of the posterior ectosylvian sulcus and the posterior sylvian gyrus.

The anterior sylvian gyrus, AIII, the anterior ectosylvian gyrus (Ea) and the banks of the rostal part of the suprasylvian sulcus (SF) do not appear to receive direct fibres from the parvicellular MGB. Degeneration appears in these areas only when the magnocellular division of the MGB is destroyed (cat GSPCAU 16, Fig. 12). *The dorsal nucleus of the MGB projects preferentially to AII (experiments of group B) and the pars lateralis of the ventral nucleus to AI (experiments of group C).*

Groups D, E and F of the experiments described show that the pars ovoidea, the nucleus ventrolateralis and the marginal zone project to the posterior ectosylvian and to the posterior sylvian gyri as well as to the cortex folded in the

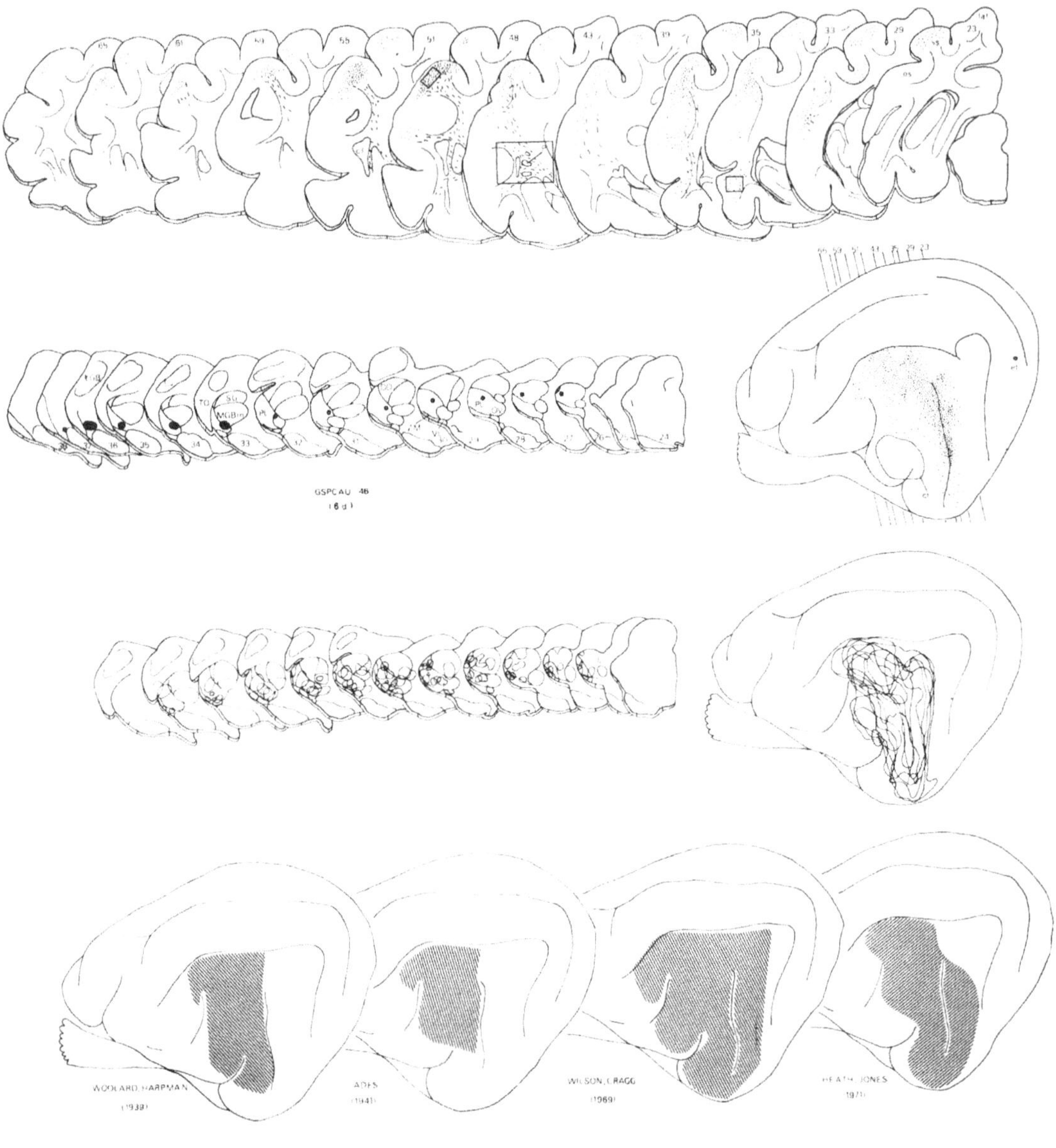

Fig. 17. The two upper rows of drawings show the lesion and the degeneration in the experiment of group G. The frames in cortical sections No. 48, 43 and 33 show the places where the photomicrographs of Fig. 7, 4 and 8, respectively, were obtained. The third row shows schematically and superimposed all the lesions and their cortical projection areas determined in the present study. The four drawings in the bottom row show the cortical projection areas of the parvicellular MGB as determined by the other authors who used anterograde degeneration methods

posterior ectosylvian sulcus. Even if it appears that there is considerable overlap among these projections, it is possible that each of these parts of the MGB has its preferential projection to a particular cortical area: Pars ovoidea projects to the

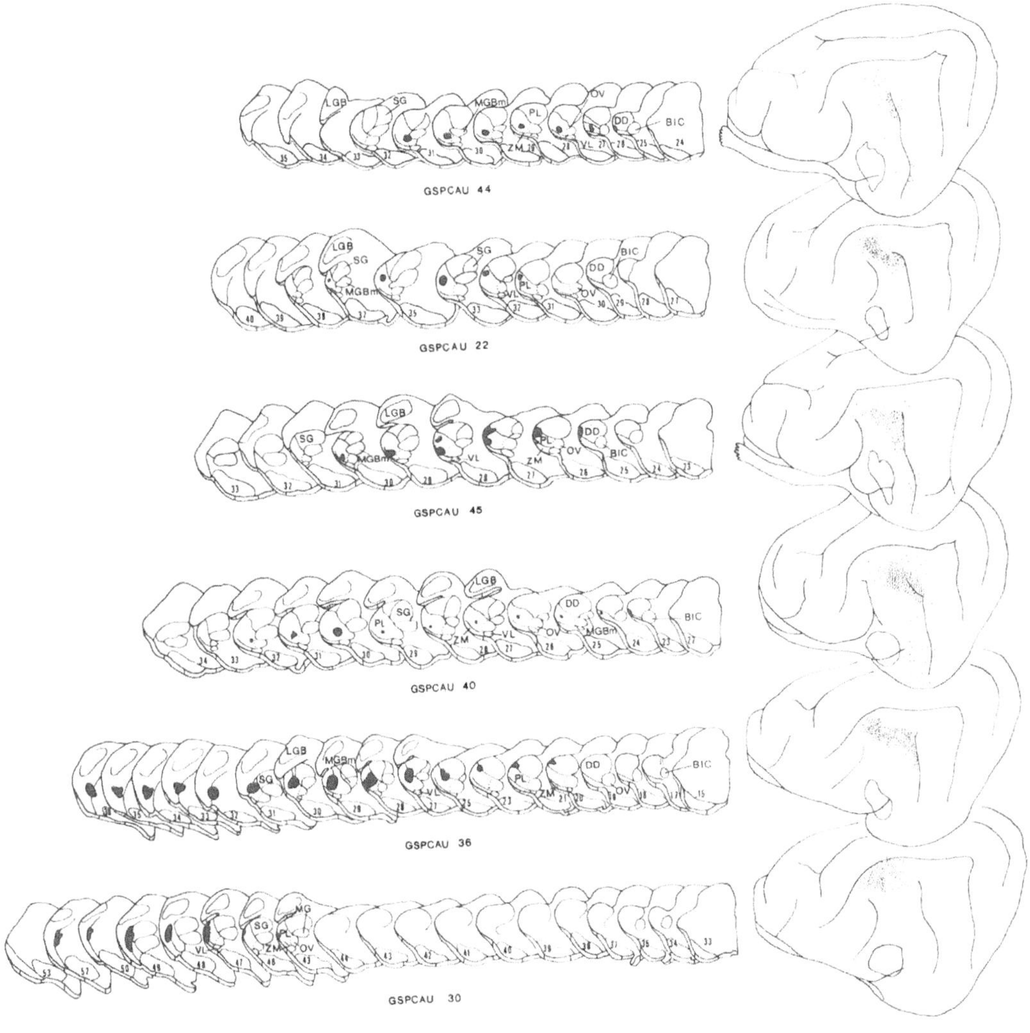

Fig. 18. Drawings of the lesions (left) and corresponding cortical areas of degeneration (right) of animals where the lateral part of the ventral nucleus of the MGB has been partially destroyed

dorsorostral part of the posterior ectosylvian gyrus; the ventrolateral nucleus to the posterior sylvian gyrus caudal to the claustrum; and the marginal zone to the ventral posterior ectosylvian gyrus.

Group C, as well as all the other experiments where the lesion involved the pars lateralis (Fig. 18), suggest the presence of topical organization of the projection to AI. As can be seen in Fig. 18 it is obvious that the caudal lesions of the pars lateralis caused degeneration in ventrocaudal areas of AI, while following more rostral lesions of this MGB subdivision the degeneration is found in more rostral parts of AI.

Discussion

Degeneration of MGB Efferent Fibres: Morphology, Time Course and Trajectory

The course of the geniculo-cortical fibres has long been known (Cajal, 1902). From the MGB some of them pass across the rostral parts of the nucleus, others run along its medial surface. This has been confirmed in the present study. The fibre course across and around the putamen and in the white matter of the hemisphere was found to be exactly similar to that described by Woollard and Harpman (1939) in the cat, and corresponding to that described by Polyak (1932) in primates and by Tunturi (1970) in the dog. It is worthy of notice that many fibres ascending to the dorsalmost part of the auditory cortex (AI) pass parallel to the cortical surface, immediately subcortically in the white matter of the middle sylvian and ectosylvian gyri. This has been noted by Woollard and Harpman (1939). and appears to be the case also in the dog (Tunturi, 1970, his Fig. 3). This feature must be taken into account in studies of the degeneration following cortical lesions in this area, but has been often overlooked. The oblique intracortical trajectory and termination in layers III and IV observed in the present study agrees with classical views (Cajal, 1911) on the termination of specific cortical afferents, and has been confirmed by electron microscopy (Sousa-Pinto and Vaaland, *in preparation*). The geniculo-cortical fibres include a strong component to layer I, as described also by Wilson and Cragg (1969). In the present study it has further been found that most of these so-called nonspecific afferent fibres pass through the cortex of the fundus of the dorsal posterior ectosylvian sulcus and spread tangentially in layer I to the neighbouring cortical surface.

The geniculo-cortical fibres passing above and across the putamen to the dorsal cortical areas were found to be thicker than those passing below the putamen to the ventral cortical areas. This may suggest that the latter are collaterals of the first, given off close to the medial edge of the putamen. The Golgi studies of Cajal (1902) and Morest (1964) show that the MGB efferents apparently are not of uniform thickness. These authors describe the axons of the projecting cells of the pars lateralis of the ventral nucleus as "très épais" and "robust" while the efferent axons of the other subnuclei are described with no qualification. In the present experiments none of the lesions produced degeneration restricted to the thin ventral fibres. On the other hand, in several experiments (groups B and C) isolated degeneration was observed of the thick and medium caliber fibres of the dorsal component passing to AI and AII. This suggest that the thinner fibres passing to the ventral areas are not collaterals of fibres from the dorsal nucleus or pars lateralis of the ventral nucleus. They might, therefore, by collaterals of fibres originating in the pars ovoidea. The findings in the experiment GSPCAU 43 (Fig. 15) are particularly suggestive since there was scanty degeneration in rostral AI and AII, but there was heavy degeneration of thick fibres to the dorsal posterior ectosylvian gyrus together with many thin degenerating fibres in the ventral parts of the auditory cortex. An alternative interpretation is that the axons of the projecting cells of the zona marginalis and of the ventrolateral nucleus, which pass mainly to the ventral auditory areas, are thinner than those of the other parts of the MGB. However, conclusive evidence is as yet lacking. Golgi studies of the MGB efferents at the medial edge of the putamen may clarify this point.

The time course of the degeneration of the MGB efferents is similar to that found in many other fibre systems. Since in the animals with very short survival times (3, 4 and 5 days) the degeneration found was similar to that of the others, it seems safe to conclude that no impregnation of retrograde degenerating fibres occurred (see Rinvik, 1968b; Wong-Riley, 1972). However, a survival period of 6 to 8 days is apparently best suited to obtain the selective silver impregnation of most of the degenerating fibres. It is interesting to note that in the animals that survived 6 to 8 days many small silver droplets were found in cortical areas showing degeneration (Fig. 10). These silver droplets correspond, at least in part, to degenerating axon terminals (see Walberg, 1971, 1972). Therefore, although this silver debris as such can not be regarded as demonstrating the existence of degenerating boutons, its presence in areas where characteristically degenerating fibres are found may be a useful indication in the search for the latter structures by electron microscopy.

The Cortical Projection Area of the MGB

The cortical projection of the MGB in the cat has been studied with the Marchi method by Woollard and Harpman (1939) and Ades (1941). In both studies the cortical projection area (Fig. 17) was found between the anterior and the posterior ectosylvian sulci, extending dorsalwards to the suprasylvian sulcus. Ventrally, the boundaries differ. Woollard and Harpman (1939) state that the area extends to a short distance above the rhinal fissure, including large areas of the anterior and posterior sylvian gyrus, while Ades (1941) marks the ventral border at the dorsal end of the pseudosylvian sulcus (Fig. 17). Wilson and Cragg (1969), using the Nauta-Gygax method, found a more extensive projection than that of Woollard and Harpman. Ventrally it extended to the rhinal fissure, and dorsally it included the ventral (lateral) bank of the suprasylvian sulcus, rostrally the anterior ectosylvian gyrus and caudally the posterior ectosylvian gyrus as well (Fig. 17). Recently, Heath and Jones (1971) presented one experiment with a lesion of the parvicellular MGB showing a cortical projection similar to that of Woollard and Harpman (1939) excluding, however, the anterior sylvian gyrus, and including most of the posterior ectosylvian gyrus (Fig. 17).

The present results agree with the studies mentioned above in showing that the most important projections of the parvicellular MGB pass to the rostral middle ectosylvian gyrus (AI) and to the middle sylvian gyrus dorsal to the tip of the pseudosylvian sulcus (AII). Dorsally and rostrally the MGB projection area was found in the present study to be bordered by the suprasylvian and the anterior ectosylvian sulci, in agreement with Woollard and Harpman (1939), Ades (1941) and Heath and Jones (1971), but not with Wilson and Cragg (1969). The latter authors included the anterior ectosylvian gyrus and the ventral bank of the suprasylvian gyrus in the projection area of the pervicellular MGB. They based this conclusion on experiments (their brains 5L, 5R and 3L) where the thalamic lesions, most probably, involved parts of the suprageniculate nucleus or of the magnocellular MGB or fibres arising from these nuclei, which are now known to project to those cortical areas (Graybiel, 1970, 1971, 1973; Heath and Jones, 1971; my experiment GSPCAU 16, Fig. 12).

Ventrally, the parvicellular MGB projection was here found not to include the insular cortex (anterior sylvian gyrus), and to include the posterior sylvian gyrus until shortly above the rhinal fissure. This observation agrees with the experiment of Heath and Jones (1971). A projection from the parvicellular MGB to the insular cortex was found by Woollard and Harpman (1939) and Wilson and Cragg (1969) but could not be confirmed in the present study or in those of Heath and Jones (1971) and Ades (1941). The explanation forwarded above is also pertinent here since in both the former studies the lesions that caused degeneration in the insular cortex involved nuclei which project to this cortical area, as shown by Heath and Jones (1971) and in my experiment GSPCAU 16 (Fig, 12). The inclusion of the posterior sylvian gyrus in the cortical projection area of the parvicellular MGB agrees with all previous studies.

The geniculate projection to the posterior ectosylvian gyrus was not found by Woollard and Harpman (1939) and Ades (1941). This may, perhaps, be ascribed to some deficiency of the Marchi method in this area, since this projection was evident in the present study, as well as in all the others made with axon stains. On the other hand, Heath and Jones found this projection extending through the whole thickness of the posterior ectosylvian gyrus, while in the present study it was found restricted to its rostral half. (Wilson and Cragg did not study the caudal half of this gyrus.) No explanation of this discrepancy can be given.

It emerges from the present and other anterograde degeneration studies, that it appears now fairly well established that the dorsal and ventral nuclei of the MGB project to a cortical area including: a) the middle ectosylvian gyrus rostral to the tip of the ectosylvian sulcus; b) the middle and posterior slyvian gyrus and c) the posterior ectosylvian gyrus (Fig. 17). The rostrocaudal extent of the projection in the latter gyrus needs confirmation. The present results stress the importance of the MGB projections to the cortex folded in the posterior ectosylvian sulcus.

The studies of Graybiel (1970, 1973) and Heath and Jones (1971) as well as my experiment GSPCAU 16 demonstrate that the magnocellular MGB (together with the suprageniculate nucleus) projects to a cortical band encircling rostrally, dorsally, and in part caudally, the projections of the principal MGB, and has also heavy projections to the cortex folded in the anterior ectosylvian and rostral suprasylvian sulci.

Comparison of the results of these anterograde degeneration studies with those of retrograde degeneration meets with considerable difficulties. These will be discussed in detail under the next heading. Suffice it to mention here that Waller (1939), Rose and Woolsey (1958), Thompson (1960), and Sharlock *et al.* (1965), among others, reported experiments where total degeneration of the MGB was found after cortical lesions. Some of the lesions used by these authors are rather large and, in addition, in all of them there was extensive involvement of the white matter. However, even the smallest lesions where complete MGB degeneration was obtained (Waller's cat No. 124 L) includes all of the cortical areas mentioned above.

The cortical projection of the MGB as delimited in the present study agrees with the cortical auditory areas as defined by electrophysiological experiments. The overall extension and situation of the principal areas (AI, AII, AIII, Ep,

Ea, SF, I and T) coincide mostly with the cortical projection areas of the dorsal, ventral, and magnocellular MGB, as defined in the present and in Heath and Jones' studies. This concordance is evident when one compares (Figs. 1A and 17).

The Cortical Projections of the Subdivisions of the MGB and the Question of "Sustaining Projections"

The present results show that each of the MGB subdivisions projects to a separate cortical area and that these areas largely coincide with the cortical auditory areas as defined electrophysiologically. This close correlation of thalamic subdivisions with functionally defined cortical areas has not yet been recognized in corresponding detail in other thalamo-cortical systems, and apparently, it has been possible to determine it here due to the thalamic subdivisions established by Morest (1964) on the basis of very detailed Golgi studies of the MGB. These observations strongly emphasize the need of similar Golgi studies of other thalamic nuclei, since it is apparent that the thalamic subdivisions established on this basis may have closer correlations with the pattern of thalamo-cortical projections than those based on classical cyto-and mylo-architectural studies.

Many studies of the medial geniculate projections to the subdivisions of the auditory cortex were made by examinations of retrograde degenerative changes in the MGB after cortical ablations. For the purpose of this discussion it appears profitable to separate among those studies the first (Waller, 1939), where acute retrograde changes (*primäre Reizung*) were used, from the others, where late retrograde changes were studied. The latter studies were started in this area by Rose and Woolsey (1949) who introduced the concept of "essential" and "sustaining" projections in order to explain the results obtained. These authors defined an essential projection as one passing to a cortical area whose destruction provokes marked degenerative changes in its related thalamic nucleus. On the other hand, sustaining projections are those passing to two (or more) cortical areas, when separate ablation of any one of these leads to no or only to "slight alterations" in the thalamus, while simultaneous destruction of both areas provokes "profound degeneration" in the related thalamic nuclei (Rose and Woolsey, 1958). The real nature of the sustaining projection has never been demonstrated and Rose and Woolsey's concept has so far remained an hypothesis. It rests on two assumptions: On one hand the existence of axon collaterals, and on the other that the thalamic cells degenerate only when all the collaterals of their axons are severed. The latter assumption agrees with the results of Fry and Cowan (1972) in the lateral mammillary nucleus of the cat but is at variance with the findings of Torvik and Brodal (1957) in the pontine reticular formation of the kitten (see Brodal, *in press*, for a detailed discussion of this point).

The original findings of Rose and Woolsey were confirmed and extended in a number of studies where the same method was used (Neff *et al.*, 1956; Diamond *et al.*, 1958; Rose and Woolsey, 1958; Neff and Diamond, 1958; Thompson, 1960; Locke, 1961; Sharlock, 1965; and others). Their results may be summarized as follows (Woolsey, 1960).

Ablation of AI causes marked degeneration restricted to the rostral $^2/_3$ of the MGB and, consequently, AI is said to receive an essential projection from that thalamic part. Lesions of AII and Ep provoke only "marginal alterations" in the MGB but, when a lesion of any of these areas involves also AI, the degeneration in the rostral $^2/_3$ of the MGB is much more intense than after isolated lesions of AI. It is concluded that the rostral $^2/_3$ of the MGB have an essential projection to AI and sustaining projections to AII and Ep. Ablation of the insulo-temporal cortex causes evident degeneration in the caudal pole of the parvicellular MGB. Degeneration is much more marked in the caudal MGB when the lesion involves AII as well, and extends to its magnocellular part when there is a simultaneous lesion of AI. This implies that the caudal pole of the parvicellular MGB has essential projections upon the I-T cortex and sustaining projections also upon AII and Ep, while the magnocellular nucleus has sustaining projections "at least to AI and probably to AII and Ep as well, but certainly also to the temporal and insular fields" (Woolsey, 1960), and to the vestibular cortex, as found later by Locke (1961).

Let us consider now the results of the present and other anterograde degeneration studies (Diamond *et al.*, 1969; Heath and Jones, 1971). These results are summarized in Fig. 19, and only the most relevant points will be mentioned here. The lateral part of the ventral nucleus of the MGB (PL) projects upon AI and receives projections from AI, AII and Ep. The dorsal nucleus (DD), which includes the caudal pole, projects to AII and receives sparse projections from AII (cp. cat 200 with 203 and 205 in Diamond *et al.*, 1969), and dense projections from the sylvian gyrus (cat 221 of Diamond *et al.*). The magnocellular nucleus projects upon AIII, the insular cortex, Ea and SF and receives projections from AI, AII, Ep, the insular and the temporal cortex. The pars ovoidea, ventrolateral nucleus and zona marginalis project upon Ep, the posterior sylvian gyrus and the cortex folded in the posterior ectosylvian sulcus, but cortical projections upon them were not detailed by Diamond *et al.* except for the ventrolateral nucleus which appears to receive a projection from the posterior sylvian gyrus (cp. cat 203 with 205 and 221 of Diamond *et al.*). It might be noted that these results agree with the so-called principle of the reciprocity of thalamocortical connections.

It is known that the thalamic nuclei will show degenerative changes when their efferent axons are cut. It is most likely that the late degenerative changes (atrophy, cell loss, gliosis) will be more marked if the afferents are destroyed together with the efferents, since, in this case, transneuronal effects will be added, and if the deafferention is complete the interneurons of the nucleus will probably degenerate as well. In this light, the information on the geniculo-cortical interconnections, made available through silver studies of anterograde axon degeneration and summarized in Fig. 19, permits, apparently, to explain the results obtained in the studies of the late retrograde changes mentioned above without resorting to the sustaining projection hypothesis. Two examples will illustrate this.

The lateral part of the ventral nucleus (which roughly corresponds to Rose and Woolsey's rostral $^2/_3$ of the nucleus) projects upon AI and receives a heavy projection from the same area. The marked degenerative changes in this part of the MGB after a lesion of AI are explained by its neurons being both deafferented and having their axons cut. This part of MGB does not project upon AII or Ep,

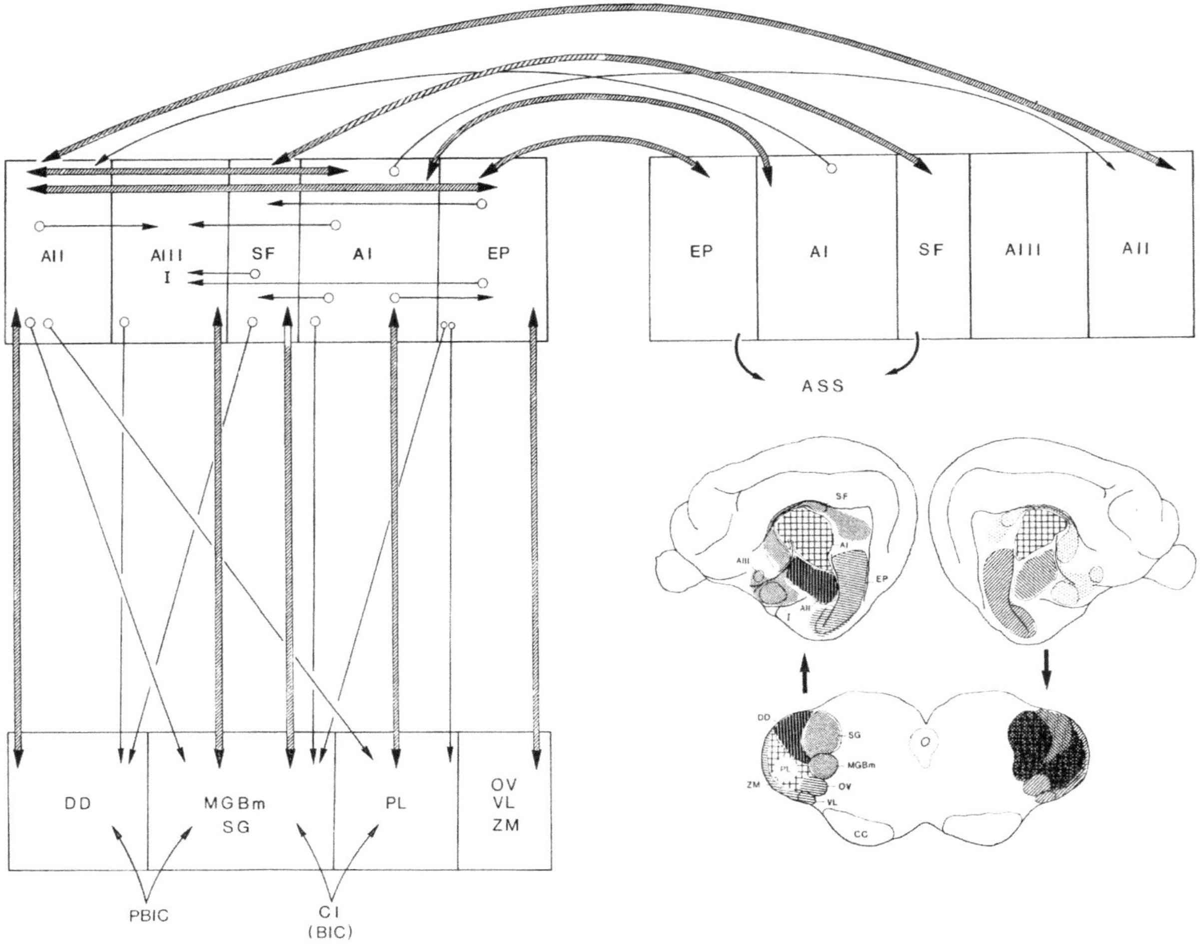

Fig. 19. Diagram of the connections of the cat auditory cortex as revealed by this and other silver impregnation studies of anterograde degeneration. On the left and on the top of the figure all connections are shown schematically. The association and commissural interconnections of the auditory cortex are shown as demonstrated by Diamond *et al.* (1968a, b). Below, to the right, maps of the interconnected parts of the auditory cortex and of the MGB in a less diagramatic manner. Connected areas are indicated by identical symbols. Geniculo-cortical projections are shown as demonstrated by Heath and Jones (1971, dotted areas) and in the present study (other areas). Cortico-geniculate projections are shown as found by Diamond *et al.* (1969)

but it receives projections from these cortical areas. Consequently, its deafferentation will be more complete and transneuronal changes will be more marked if AII and Ep are destroyed together with AI. On the other hand, if only AII or Ep are destroyed the consequent transneuronal degeneration in the nuclear subdivision may be minimal or inapparent, since its most important afferent and efferent connections (with AI) are preserved. This is exactly what was observed in the late retrograde degeneration studies referred to above.

The magnocellular MGB projects upon the insular cortex, to Ea and to SF but does not project to AI, AII or Ep. However, marked retrograde changes in that nucleus are found only when AI is destroyed together with the insular cortex. This might be explained by the fact that AI has a strong projection to the magnocellular nucleus and, in addition, AII and Ep also project upon it (Diamond *et al.*, 1969). The persistance of these afferents, together with the preservation of its efferents to Ea and SF are apparently enough to prevent the appearance of marked retrograde changes in the nucleus, after the destruction of one its cortical target areas, which Graybiel (1970, 1971, 1973) and Heath and Jones (1971) have shown to be the insular cortex.

The findings interpreted as late retrograde changes in the MGB after cortical lesions appear, therefore, in part to be due to transneuronal influences. This is suggested also by Brodal (*in press*). This interpretation holds, apparently, for every cortical or MGB subdivision, as will be obvious if one compares the scheme of Fig. 19 with the results of the retrograde degeneration studies.

This interpretation, probably, does not apply to the studies of acute retrograde changes (*primäre Reizung*). However, only Waller (1939) has studied the geniculocortical projections using that method. A comparison of his results with those of the present and the other anterograde degeneration studies is hampered by the lack of details in his descriptions concerning the subnuclei of the MGB as well as by the extension of Waller's lesions to several of the cortical subdivisions as defined at present. However, if one selects from his material cases with lesions apparently restricted to particular subdivisions of the auditory cortex and compares them with the present results, there appears to be a fair degreee of agreement. Waller's cats No. 123, 124R and 125 which have lesions restricted to AI show acute retrograde changes restricted to an area which appears to be the pars lateralis of the ventral nucleus. His experiments No. 126, with a lesion restricted to AII shows changes apparently restricted to the dorsal nucleus of the MGB. His experiment 127R where a complex lesion involves parts of the insular cortex, of AII, and of the dorsal Ep, is the only one where the posterior sylvian gyrus and the ventral Ep were destroyed. In the MGB degeneration was found in different parts, but this is also the only case where degeneration was found in the ventrolateral nucleus and in the marginal zone.

Another point in the present results requires discussion. The part of AI rostral to the dorsal tip of the posterior ectosylvian sulcus receives many projection fibres from the pars lateralis of the MGB. In the present study fibre degeneration was found in the low frequency region of Woolsey's (1960) AI, which is caudal to that sulcus, only when the lesions involved the pars ovoidea. This appears to support the inclusion of the latter cortical area, as well as of the banks of the related sulcus, in AI, since the pars lateralis and the pars ovoidea are cytologically similar and share the same ascending afferent connections (Morest, 1964, 1965b; Noort, 1969).

However, some differences seem to exist not only between pars lateralis and pars ovoidea, but also between their cortical projection areas. Pars lateralis has a straight lamellar arrangement (Morest, 1965a), receives descending projections from AI, AII and EP (Diamond *et al.*, 1969) and projects upon the rostral part of AI (group C, Fig. 13). On the other hand, pars ovoidea has a coiled arrangement of its laminae (Morest, 1965a), receives descending projections only from Ep (Diamond *et al.*, 1969) and, in addition to its projection to the posterior low-frequency region of Woolsey's AI, sends thin fibres to Ep and to the banks of the ventral part of the posterior ectosylvian sulcus (group D, Fig. 14). In addition, the part of Woolsey's AI caudal to the posterior ectosylvian sulcus, as well as the caudal bank of this sulcus, have a cytoarchitecture similar to Ep, which is different from the rostral part of AI (Rose, 1949, Sousa-Pinto, 1973). Like Ep, this cortical area has projections to the deep layers of the superior colliculus, while AI has no projections to this area (Paula-Barbosa and Sousa-Pinto, 1973). All these differences appear to suggest that the caudal bank of the dorsal part of the posterior ectosylvian sulcus as well as the neighbouring area of the convexity of the posterior ectosylvian gyrus may be more closely related to Ep than to AI. Electrophysiological studies of the very extensive cortex folded in that sulcus are necessary to decide the apparent conflict with the evidence brought forward by Woolsey (1960) in favour of the extension of AI to this cortical area.

The Organization of the Thalamo-Cortical Auditory System

From the preceeding discussion it appears that the MGB projects upon the cortex in a fairly organized fashion. The larger of Morest's (1964) subdivisions of the nucleus have independent cortical projection areas, which largely coincide with the electrophysiologically demonstrated auditory areas and show very little mutual overlap. The smaller subdivisions appear to project to a common cortical area where, however, preferential regions may be segregated for each one of the small subnuclei. The auditory cortical areas project to the MGB subdivisions in a reciprocal manner. It interhemispheric and ipsilateral interconnections have recently been mapped (Diamond *et al.*, 1968a, b). Fig. 19 shows, very schematically, the information available at present on the connections of the thalamocortical auditory system of the cat. The diagram is based on the present and other recent anterograde degeneration studies where the sensitive silver methods have been used. Even if the diagram may be oversimplified, it is evident that many interconnections are established among the different subdivisions of this system.

The circuit diagram of Fig. 19 suggests that the thalamocortical auditory system has groups of nuclei with evident differences in their pattern of connectivity. Pars lateralis of the ventral nucleus of the MGB and AI have strong reciprocal interconnections. This MGB subdivision receives the bulk of the ascending auditory pathway (Rasmussen, 1964; Moore and Goldberg, 1963; Powell and Hatton, 1969; Noort, 1969) and has cytoarchitectonic features indicative of being a modality specific area (Morest, 1964). In addition to the ascending connections, the only other established connections of pars lateralis are with AI, AII and Ep (Fig. 19).

AI, like its thalamic correspondent (PL), appears to have rather specific afferent connections. In addition to fibres from the pars lateralis, it receives projections from the ipsilateral AII and from the contralateral AI (Fig. 19). All these afferent projections appear to be specific and topically organized (Diamond *et al.*, 1968a, b; Rose and Woolsey, 1949; the present study, Fig. 18). A tonotopic organization of AI is suggested by Woolsey and Walzl's (1942) study of its connections with the cochlea, and the present and Rose and Woolsey's (1949) results support this view.

In agreement with the apparent specificity suggested by its connections, AI shows cytoarchitectonic features characteristic of a specific sensory cortical area (Rose, 1949). It is the only cortical area activated by click stimuli under deep barbiturate anesthesia (Bremer and Dow, 1939) and, apparently, is not activated by other, viz. visual or somesthetic, stimuli (Galambos, 1960; Stewart and Starr, 1970).

Therefore, it appears that *the pars lateralis-AI system may be the modality specific and spatially organized core of the thalamo-cortical auditory system.* It has few and specific afferent fibre systems and through its many efferents it may influence bilaterally most of the auditory cortex (Fig. 19).

The analysis of the diagram shows a second system, including the suprageniculate-magnocellular complex and its cortical target, the insular cortex, SF and AIII, which, with regard to its connections, contrasts with the pars lateralis-AI system. Its thalamic part receives auditory projection fibres from the inferior colliculus (Rasmussen, 1964; Moore and Goldberg, 1963; Powell and Hatton, 1969; Noort, 1969) and from a parabrachial pathway (Morest, 1965b). But the suprageniculate-magno cellular complex receives also spinal projections (Whitlock and Perl, 1959) and is activated by auditory, somatic and vibratory impulses (Poggio and Mountcastle, 1960). It receives fibres from all subdivisions of the auditory cortex (Diamond *et al.*, 1969) as well as from the somatic sensory cortex (Jones and Powell, 1968; Rinvik, 1968, b). Consequently, it is a region with many afferent connections conveying different modalities of information. It has restricted projections to AIII, to the insular cortex and to the suprasylvian fringe (Graybiel, 1970, 1971, 1973; Heath and Jones, 1971). Its cortical target area, in addition to its reciprocal thalamic connections, receives fibres from all the ipsilateral auditory areas (Diamond *et al.*, 1968a, b, see Fig. 19). Contrasting with its numerous and unspecific afferent connections, this cortical area appears to have very restricted efferent connections. In addition to those to the magnocellular MGB, its only demonstrated efferents pass to the dorsal nucleus of the MGB (Fig. 19) and to the suprasylvian association area of Woosley (1960).

This second system, including the suprageniculate-magnocellular complex, AIII, the insular cortex and the suprasylvian fringe, appears as a system with wide and unspecific input where integration of several modalities may occur.

Summary

The cortical projections of the medial geniculate body were experimentally studied in 21 adult cats, using several silver impregnation methods. The degeneration occasioned by stereotaxic lesions of the medial geniculate body was mapped using a precise method for reference to the cortical surface. Relatively uncomplicated lesions of each of the several subdivisions (Morest's, 1964) of the parvicellular MGB were obtained.

The cortical projection area of the parvicellular MGB was found to include: a) the middle ectosylvian gyrus rostral to the tip of the posterior ectosylvian sulcus (AI); b) the transverse (AII) and posterior (T) sylvian gyrus; c) the rostral part of the posterior ectosylvian gyrus (Ep) and d) the bottom and both banks of the posterior ectosylvian sulcus. The experiments with lesions restricted to the lateral part of the ventral nucleus of the MGB demonstrated that the projection of this subdivision is restricted to AI. Experiments with lesions of the dorsal nucleus (including the caudal part of the nucleus) have shown that this subdivision projects to AII. The experiments with lesions that included the pars ovoidea, the zona marginalis, or the ventrolateral nucleus, demonstrated that these small subdividions project to Ep and to the bottom and banks of the posterior ectosylvian sulcus. The latter experiments suggest, in addition, that each one of the three small subdivisions project preferentially to the dorsal part of Ep, to the ventral part of Ep and to the posterior sylvian gyrus, respectively.

On the basis of these observations, the importance of detailed Golgi studies of the thalamus is emphasized. It appears very likely that the subdivisions of the thalamus as shown by the Golgi methods may be more closely connected with individual cortical areas than are the thalamic subdivisions shown by the classical cytoarchitectural studies.

Correlation of the present results with the corticogeniculate projections (Diamond *et al.*, 1969), and with the cortical projections of the magnocellular MGB (Graybiel, 1970, 1971, 1973; Heath and Jones, 1971), suggests that the hypothesis of essential and sustaining projections is not necessary in order to interpret the results that Rose and Woolsey (1949) obtained in their cellular degeneration study of the cortical projections of the medial geniculate body. The cellular changes found in the latter nucleus more than two weeks after cortical lesions are, most probably, due do anterograde transneuronal effects as well as to retrograde effects.

The findings are discussed in relation to pertinent anatomical and physiological studies. It seems that the thalamo-cortical auditory system includes two groups of nuclei which have different connections and presumably different functions. The first, apparently, is modality specific and spatially organized. It includes AI and the lateral part of the ventral nucleus of the medial geniculate body. The second group of nuclei includes AIII, SF, the insular cortex, and the suprageniculate-magnocellular complex.

Acknowledgements. The author thanks Professor Alf Brodal, Anatomical Institute, University of Oslo, Norway, for helpful discussions and criticisms, as well as for correcting the manuscript.

The histological work made by Mrs. Manuela Pacheco and the secretarial help of Miss M. Emília Mendes are gratefully acknowledged.

This research received financial support from the Instituto de Alta Cultura (Project PMC/5) as well as from NATO (Research Grant 532).

References

Ades, H. W.: Connections of the medial geniculate body in the cat. Arch. Neurol. Psychiat. (Chic.) **45**, 138–144 (1941)

Bremer, F., Dow, R. S.: The cerebral acoustic area of the cat. A combined oscillographic and cytoarchitectonic study. J. Neurophysiol. **2**, 308–318 (1939)

Brodal, A.: Anterograde and retrograde degeneration of nerve cells in the central nervous system. In: Histology and general pathology of the human nervous system, eds. W. Haymaker and R. D. Adams, in press.

Cajal, S. R.: Estrutura del tuberculo cuadrigémino posterior, cuerpo geniculado interno y vias acústicas centrales. Trab. Lab. Invest. Biol. **1**, 207–227 (1902)

Cajal, S. R.: Histologie du système nerveux de l'homme et des vertébres. Paris: Maloine 1911 (ed. of the Cons. Sup. Inv. Cient., Madrid, 1972)

Diamond, I. T., Chow, K. L., Neff, W. D.: Degeneration of the caudal medial geniculate body following cortical lesion ventral to auditory area II in the cat. J. comp. Neurol. **109**, 349–362 (1958)

Diamond, I. T., Jones, E. G., Powell, T. P. S.: Interhemispheric fiber connections of the auditory cortex of the cat. Brain Res. **11**, 177–193 (1968a)

Diamond, I. T., Jones, E. G., Powell, T. P. S.: The association connections of the auditory cortex of the cat. Brain Res. **11**, 560–579 (1968b)

Diamond, I. T., Jones, E. G., Powell, T. P. S.: The projection of the auditory cortex upon the diencephalon and brain stem in the cat. Brain Res. **15**, 305–340 (1969)

Eager, R. P.: Selective staining of degenerating axons in the central nervous system by a simplified silver method: spinal cord projections to external cuneate and inferior olivary nuclei in the cat. Brain Res. **22**, 137–141 (1970)

Fink, R. P., Heimer, L.: Two methods for selective silver impregnation of degenerating axons and their synaptic endings in the central nervous system. Brain Res. **4**, 369–374 (1967)

Fry, F. J., Cowan, W. M.: A study of retrograde cell degeneration in the lateral mammillary nucleus of the cat, with special reference to the role of axonal branching in the preservation of the cell. J. comp. Neurol. **144**, 1–24 (1972)

Galambos, R.: Discussion. In: Neural mechanisms of the auditory and vestibular systems, p. 218, eds. G. L. Rasmussen and W. F. Windle, Springfield: Thomas 1960

Grant, G., Aldskogius, H.: Silver impregnation of degenerating dendrites, cells and axons central to axonal transection. I. A Nauta study on the hypoglossal nerve in kittens. Exp. Brain Res. **3**, 150–162 (1967)

Graybiel, A. M.: Some thalamocortical projections of the pulvinar-posterior system of the thalamus in the cat: Brain Res. **22**, 131–136 (1970)

Graybiel, A. M.: Some fiber connections of the posterior thalamus in the cat. Doctoral dissertation, Massachussets Institute of Technology. Cambridge 1971

Graybiel, A. M.: The thalamo-cortical projection of the so-called posterior nuclear group: a study with anterograde degeneration methods in the cat. Brain Res. **49**, 229–244 (1973)

Heath, C. J., Jones, E. G.: An experimental study of ascending connections from the posterior group of thalamic nuclei in the cat. J. comp. Neurol. **141**, 397–426 (1971)

Jones, E. G., Powell, T. P. S.: The projection of the somatic sensory cortex upon the thalamus in the cat. Brain Res. **10**, 369–391 (1968)

Locke, S.: The projection of the magnocellular medial geniculate body. J. comp. Neurol. **116**, 179–193 (1961)

Monakow, C.: Weitere Mitteilungen über einige durch Exstirpation circumscripter Hirnrindenregionen bedingte Entwicklungshemmungen des Kaninchengehirns. Arch. Psychiat. Nervenkr. **12**, 535–544 (1882)

Moore, R. Y., Goldberg, J. M.: Ascending projections of the inferior colliculus in the cat. J. comp. Neurol. **121**, 109–136 (1963)

Morest, D. K.: The neuronal architecture of the medial geniculate body of the cat. J. Anat. (Lond.) **98**, 611–680 (1964)

Morest, D. K.: The laminar structure of the medial geniculate body of the cat. J. Anat. (Lond.) **99**, 143–160 (1965a)

Morest, D. K.: The lateral tegmental system of the midbrain and the medial geniculate body: study with Golgi and Nauta methods in cat. J. Anat. (Lond.) **99**, 611–634 (1965b)
Nauta, W. J. H.: Silver impregnation of degenerating axons. In: New research techniques of neuroanatomy, p. 17–26, ed. W. F. Windle. Springfield: Thomas 1957
Neff, W. D., Diamond, I. T.: The neural basis of the auditory discrimination. In: Biological and biochemical basis of behaviour, p. 101–126, eds. H. F. Harlow and C. N. Woolsey. Madison: University of Wisconsin Press 1958
Neff, W. D., Fisher, J. F., Diamond, I. T., Yela, M.: Role of auditory cortex in descrimination requiring localization of sound in space. J. Neurophysiol. **19**, 500–512 (1956)
Noort, J. van: The structure and connections of the inferior colliculus. Amsterdam: Van Gorcum & Co. N. V. 1969
Paula-Barbosa, M., Sousa-Pinto, A.: Auditory cortical projections to the superior colliculus in the cat. Brain Res. **50**, 47–61 (1973)
Poggio, G. F., Mountcastle, V. B.: A study of the functional contributions of the lemniscal and spinothalmic system to somatic sensibility. Central nervous mechanisms in pain. Bull Johns Hopk. Hosp. **106**, 266–316 (1960)
Polyak, S.: The main afferent fiber systems of the cerebral cortex in primates (Auditory system). Univ. Calif. Publ. Anat. **2**, 79–104 (1932)
Powell, E. W., Hatton, J. B.: Projections of the inferior colliculus in cat. J. comp. Neurol. **136**, 183–192 (1969)
Rasmussen, G. L.: Anatomic relationship of the ascending and descending auditory systems. In: Neurological aspects of auditory and vestibular disorders, p. 5–19, eds. W. S. Fields and B. R. Alford. Springfield: Thomas 1964
Rinvik, E.: A re-evaluation of the cytoarchitecture of the ventral nuclear complex of the cat's thalamus on the basis of corticothalamic connections. Brain Res. **8**, 237–254 (1968a)
Rinvik, E.: The corticothalamic projection from the pericruciate and coronal gyri in the cat. An experimental study with silver-impregnation methods. Brain Res. **10**, 79–119 (1968b)
Rioch, D. Mck.: Studies on the diencephalon of carnivora. Part I. The nuclear configuration of the thalamus, epithalamus and hypothalamus of the dog and cat. J. comp. Neurol. **49**, 1–119 (1929)
Rose, J. E.: The cellular structure of the auditory region of the cat. J. comp. Neurol. **91**, 409–440 (1949)
Rose, J. E., Woolsey, C. N.: The relations of the thalamic connections cellular structure and evocable electrical activity in the auditory region of the cat. J. comp. Neurol. **91**, 441–466 (1949)
Rose, J. E., Woolsey, C. N.: Cortical connections and functional organization of the thalamic auditory system of the cat. In: Biological and biochemical basis of behaviour, p. 127–150, eds. M. F. Harlow and C. N. Woolsey. Madison: University of Wisconsin Press 1958
Scharlock, D. P., Neff, W. D., Strominger, N. L.: Discrimination of tone duration after bilateral ablation of cortical auditory areas. J. Neurophysiol. **28**, 673–681 (1965)
Sousa-Pinto, A.: The structure of the first auditory cortex (AI) in the cat. I. Light microscopic observations on its organization. Arch. Ital. Biol. **111**, 112–137 (1973)
Sousa-Pinto, A., Vaaland, J. L.: Experimental electron microscopic study of the specific afferents of the cat first auditory cortex. In preparation
Stewart, D. L., Starr, A.: Absence of visually influenced cells in auditory cortex of normal and congenitally deaf cats. Exp. Neurol. **28**, 525–528 (1970)
Thompson, R. F.: Function of auditory cortex of cat in frequency discrimination. J. Neurophysiol. **23**, 321–334 (1960)
Torvik, A., Brodal, A.: The origin of reticulospinal fibers in the cat. An experimental study. Anat. Rec. **128**, 113–137 (1957)
Tunturi, A. R.: The pathway from the medial geniculate body to the ectosylvian auditory cortex in the dog. J. comp. Neurol. **138**, 131–136 (1970)
Walberg, F.: Does silver impregnate normal and degenerating boutons? A study based on light and electron microscopical observation of the inferior olive. Brain Res. **31**, 47–65 (1971)

Walberg, F.: Further studies on silver impregnation of normal and degenerating boutons. A light and electron microscopical investigation of a filamentous degenerating system. Brain Res. **36**, 353–369 (1972)

Waller, W. H.: Thalamic degeneration induced by temporal lesion in the cat. J. Anat. (Lond.) **74**, 528–536 (1939)

Whitlock, D. G., Perl, E. R.: Afferent projections through the ventrolateral funiculi to the thalamus of cat. J. Neurophysiol. **22**, 133–148 (1959)

Wiitanen, J. T.: Selective silver impregnation of degenerating axons and terminals in the central nervous system of the monkey (Macaca mulatta). Brain Res. **14**, 546–548 (1969)

Wilson, M. E., Cragg, B. G.: Projections from the medial geniculate body to the cerebral cortex in the cat. Brain Res. **13**, 462–475 (1969)

Wong-Riley, M. T. T.: Changes in the dorsal lateral geniculate nucleus of the squirrel monkey after unilateral ablation of the visual cortex. J. comp. Neurol. **146**, 519–548 (1972)

Woollard, H. H., Harpman, A.: The cortical projection of the medial geniculate body. J. Neurol. Psychiat. **2**, 35–44 (1939)

Woolsey, C. N.: Organization of cortical auditory system: a review and a synthesis. In: Neural mechanisms of the auditory and vestibular systems, p. 165–180, eds. G. L. Rasmussen and W. F. Windle. Springfield: Thomas 1960

Woolsey, C. N., Walzl, E. M.: Topical projection of nerve fibers from local regions of the coclea to the cerebral cortex of the cat. Bull. Johns Hopk. Hosp. **71**, 315–344 (1942)

Subject Index

Abbreviations 6
anterograde degeneration 30, 32, 34
ascending auditory pathways 36
association auditory connexions 35
athrophic changes 7, 32
auditory cortical areas, ectosylvian anterior 7, 25, 32, 34
—, ectosylvian posterior 7, 16, 23, 25, 28, 32, 34
—, first 7, 16, 19, 21, 25, 27–29, 32, 34–36
—, insular 16, 30, 32, 34, 36
—, second 7, 16, 21, 25, 28, 29, 32, 34, 36
—, suprasylvian fringe 7, 25, 32, 34, 36
—, temporal 16, 32
—, third 7, 16, 25, 32, 36
axon collaterals 28, 31

Brachium of the inferior colliculus 9

Cell loss 7, 32
claustrum 12
cochlear representation 19
cortical, layers 10, 14, 28
— projections of medial geniculate body 7, *25*, *29*, *31*, 34
crus cerebri 12, 25
cytoarchitecture 31, 35, 36

Degenerating, boutons 10, 12, 14, 29
— passing fibres 10, 12, 14, 28
dog 28
dorsal nucleus (division, lobe) of medial geniculate body 7, 9, *14*, 16, 19, 20, 21, 23, 25, 32, 34

Efferent fibres of medial geniculate body *25*, 29
electrode track *14*
essential projections 7, *31*

Fibres, geniculo-cortical 10, *28*
fixation 8, 9

Gliosis 32
gyrus, ectosylvian 14
—, ectosylvian anterior 16, 29
—, ectosylvian middle 12, 16, 18–21, 23, 25
—, ectosylvian posterior 12, 14, 16, 19–21, 23, 25, 27, 29, 30, 34
—, lateral 14
—, suprasylvian 14, 16
—, sylvian 16
—, sylvian anterior 16, 29
—, sylvian middle 12, 14, 18, 20, 25, 29, 32
—, sylvian posterior 14, 19–21, 25, 29, 30, 32, 34

Hippocampus 14

Inferior colliculus 36
interhemispheric auditory connexions 35
internal capsule 12
interneurons 32

Lateral part of ventral nucleus of medial geniculate body (see pars lateralis) low frequency region of first auditory cortex 34

Magnocellular nucleus of medial geniculate body 7, *9*, 16, 25, 29, 30, 32, 34, 36
medial geniculate body 7, *9*, 14, 16
methods *7*
— Eager 9, 10, 12
— Fink and Heimer 9, 10
— Golgi 31
— Marchi 7, 29, 30
— Nauta 7, 9, 10
— Thionin 9
— Wiitanen 9, 10, *12*
modality specificity 36

Optic tract 12, 16
organization of auditory connexions *35*

Pars, lateralis of medial geniculate body 9, *16*, 18, *19*–23, 25, 32, 34–36
—, ovoidea of medial geniculate body 9, *19*–21, 25, 32, 35
perfusion 8
post-subicular cortex 14
pre-terminal degeneration 9, 10
primates 28
principal (parvicellular) nucleus of medial geniculate body 7, 9, 25, 29, 30, 32
putamen 12, 14, 28

Reticular complex of thalamus 18
retrograde changes 12, 29–*31*, 34

Somatic sensory cortex 36
spatial organization 36
stereotaxic lesions 7, 8

sulcus, ectosylvian anterior 16, 29
—, ectosylvian posterior 14, 16, 19–21, 23, 25, 29, 32, 35
—, pseudosylvian (sylvian) 19, 29
—, rhinal 29
—, suprasylvian 9, 14, 29
superior colliculus 35
suprageniculate nucleus 7, 10, 16, 29, 30, 36
survival time 8
sustaining projections 7, *31*

Terminal degeneration 9, 10
time course of degeneration 12, *29*
tonotopic organization 36
topic relationships 7, 19, *27*, 36
transneuronal degeneration 32, 34

Ventral division (lobe) of medial geniculate body 7, 16, 19
ventrolateral nucleus of medial geniculate body 9, *21*, 27, 28, 32, 34
vestibular cortex 32

Zona marginalis (marginal zone) of medial geniculate body *23*, 25, 27, 28, 32, 34

www.ingramcontent.com/pod-product-compliance
Ingram Content Group UK Ltd.
Pitfield, Milton Keynes, MK11 3LW, UK
UKHW021608110926
15112UKWH00017B/128

* 9 7 8 3 6 4 2 9 5 2 4 8 7 *